I0394305

BRETT WILLIAMS
BANG
GOES THAT THEORY

authorHOUSE

AuthorHouse™ UK
1663 Liberty Drive
Bloomington, IN 47403 USA
www.authorhouse.co.uk
Phone: 0800.197.4150

© 2018 Brett Williams. All rights reserved.

No part of this book may be reproduced, stored in a retrieval system, or transmitted by any means without the written permission of the author.

Published by AuthorHouse 10/09/2018

ISBN: 978-1-5462-9778-9 (sc)
ISBN: 978-1-5462-9777-2 (hc)
ISBN: 978-1-5462-9937-0 (e)

Print information available on the last page.

Any people depicted in stock imagery provided by Getty Images are models, and such images are being used for illustrative purposes only.
Certain stock imagery © Getty Images.

This book is printed on acid-free paper.

Because of the dynamic nature of the Internet, any web addresses or links contained in this book may have changed since publication and may no longer be valid. The views expressed in this work are solely those of the author and do not necessarily reflect the views of the publisher, and the publisher hereby disclaims any responsibility for them.

THE BOOKS COVER

The bible cannot be trusted. The ten main points can.

D N A what are you. A man or a mouse.
The end of the universe's expansion?
What's a black hole?
Ice age watered down.
Tides don't go in and out
Lights fast what's as fast but stronger.

The physiatrists, scientists and the professional bodies say that there is no limit to how small. Large, fast or slow that matter or elements can go. That is why I came up with there must be Something that can go faster than light and there be can be no end to its speed. Find out later.

Tilly force field. This is my wording for gravity as gravity as defined by Isaac Newton's does not fit the bill for this explanation. So what can it be?

The centre of all the planets and sun's are iron based. If this is So why is the earth's iron in the north? For now.

All the answers in this book.

The bible. Evolution which to believe.

Please note there are no chapters as I do not see the point of them why waste paper or digital space.

LET US LOOK AT THE COVER

When I say look at the cover I do not mean that you must go to all the trouble of finding the front cover. The following will give you the information to the theories put forward by the scientists and physicists. These are my way of looking logically at the universes theories.

Gravity and the Big Bang who's pulling whom.
I can light up a black hole.
D.N.A. what are you a man or a mouse.
Tides don't go in and out.
Gravity controls the universe. Not in my book.
Space and time, my watch is on the blink.
Ice age. Watered down.
Gases. Let's light them up.
Eisenstein's P.I. twenty-two over seven cannot be done. Rubbish.

The sun and black holes. Not as black as it looks. Let's shine Some light on them.

The moon. Is the moons gravity the cause of the tides? Who's pulling whom?

The moons gravity can't pull me, never mind our tides.

Ice age. Not as cold as you think.

The earth's spin. Was it knocked for a six?

The Big Bang theories

I can agree on one of their explanations about the Big Bang and that it really happened. They say that the gases in the universe in time will turn into a white dwarf and then into a black hole. I disagree. I will explain later about white dwarfs to my readers that no nothing about them. This info I got of the internet.

ITS RUBBISH

D.N.A.

Species can copulate, but the young are infertile. This basically means that evolution cannot breed a different species. Species do evolve in their own species. This means that no new species can possibly appear. So a bacteria is a bacteria and will always be bacteria.

The land was as one land mass. And it was slowly eroded until there were different continents as can be seen today. This is not possible.

MORE RUBBISH.

103 days of plague=4/5 months. This is too long for these plagues to have lasted. What is the most likely time span?

Black holes lit up. I can see into this

Dark matter; this has messed up my mind.

String theory, this has me all tied up.

Also in this book

The Bible v the Charles Darwin's theory and Bang goes the Big Bang theory? Everything you have learned So far will be turned on its head. You think the sun is a powerful force. It's hot but let's cool it down. What has a more powerful force?

The Big Bang. Came in like a lion and will go out like a lamb?

The sun. Not such a bright spark. The colours are a wash out.

The tides. Do you think they are going in and out, think again?

The universe. Is it expanding? Not from where I stand.

Evolution. Are you a man or a mouse?

Space and time. Who's getting younger? My watch is winding down. Everything you have learned So far will be turned on its head. You think the sun is a powerful force. It's hot but let's cool it down. What has a more powerful force?

The national space centre does not know if the universe will go on expanding or it will implode and reverse back to its original state. I wish these boffins would make up their minds. I have the answers. Find out later?

If the Earth's is spinning on its axis was caused by an object hitting it? What hit the sun's to make them spin? What pull does the moon really have? Ice age means grass eating animals, plants, and insects plus all plants and bacteria would have died out. The answers are scattered through the book. The answers to all the questions, scientists, physicists, and biologists only have theories for. A lot of the following is based on logic there are Some theories. If my logics are scientifically proved wrong, then they are just theories.

Not just for the scientific mind but lots of interest for those who are not inclined that way but would like to know more. There are Also bible revelations. There are So many important points in this book I hope I have put them in the correct order of their importance.

Light speed is 299,792,458, metres per second.

Milky Ways galaxies look less than Andromeda because the majority are hidden behind others I will explain why later.

Mars asteroids are small as seen from space ships So them as hand size. That's from billion of years ago.

The earth has a Solid inner core, a liquid outer core, a mantle, and a crust these are only theories.

METEORS

There will be items about meteors being burned up in our atmosphere before they land. If you have missed one, you do not must panic as there are a lot more on their way. The archaeologists have discovered cave drawings. Why are there no drawings of rainbows? I would have thought that if they did draw pictures of the animals and birds that they encounter in their environment then surly if there was a rainbow they would have included them in their artistic drawings How about that for the historians to Sort out? There are cities all over the world, in places that have no way that they have been able to communicate with each other. And if they did happen to meet the first barrier would be the different languages. There is one thing that they all have in common and that is they all must have a religious belief. You have tribes in all parts of the worlds that have a different way of worshipping their gods. It would not surprise me if there are still tribes that have not been discovered. How is it that in a lot of these places they have all managed to build pyramids? There is only one explanation and that is the bible has given the reason for this. Our Lord divided the people to different parts of the world So that they could not communicate with each other. The reason for this was that they where building towers into the sky and they were being worshiped by the people as gods. They took their knowledge of building pyramids with them. Can any of the scientists come with any other way? Maybe they were clairvoyant. Some of these people lived in places were there were no stone to build pyramids, So they improvised and made them from what they had to hand and that is why they created mounds from Soil. This city was founded nine hundred yeas A.D. These mounds contain twenty million tons of earth and it took nine million hours of work to complete and centuries to build. They had no form as in a wheel that would make the job easier. These mounds had to be constructed by hard labour. These South America Natives could grow corn all the year around and I should imagine this would be six crops a year or a continua's crop if Sown on a regular basic. This city flourished for centuries and then suddenly disappeared from the face of the earth. There were no neighbours in the near surrounds, So they had to rely on their own people. What were found below the top layer of clay were numerous bodies. They were all laid out in rows which suggested that they were all sacrificed. The reason for this was that there was a period of a drought that did not abate. So this was an offering to the gods. The bodies were all female from fifteen to twenty-five years old and the rest were children. There were 39 bodies of men that had met a violent death. No one was able to go anywhere as the drought was widespread. And this was an iSolated city. If there where no cities anywhere in the region as with lots of other places where humans could be found did they come from. And you must remember that there where no monkeys or apes in these places, that surly puts a spanner in the works for those who do not believe in a creator. The time span that this city lasted was up to thirteen hundred A.D. that meant that it survived for only two hundred and fifty years. There was no pollen found in the Soil at the end of this period which meant that there was no plant life. There was a barrier of trees around the main part of the city they reckon that twenty thousand trees were used. There were 72 mounds created. The experts in different fields have surmised that there was an uprising in the local population. I should have imagined that was because of a food shortage. I can well imagine that there was quite a lot of unrest when their females were taken as a sacrifice.

Some planets do not have a tilt. The astrologers cannot understand why this is So. They have given different reasons for this. They are looking at this from the wrong angle. You will find loads of puns in this book I hope they give you a laugh or at least Some amusement. If I can Sort this out then it could answer a lot of their stupid theories. As with most of my theory's/logics there straight to the point not, going around the bushes.

THE SPEED OF LIGHT

Einstein said there is nothing faster than light. This speed is 299,792,458, metres per second.

WRONG

Sorry I think I have found Something faster. I will let you know later. I can prove what is faster than light no theory but a scientific way of proving it. *How about that for a revelation:*

Why can we see the entire universe from Earth? You look up to the heavens and there it is the Milky Way. How come is it that the astrologers have not thought about how unusual this is? I can hear my readers that saying he's right. They talk about the age of the universe and how Andromeda galaxy is the oldest galaxy in the universe. Let us stand back to have a look at this statement. Do not stand back to far as we don't want any part of it not to be visible. This statement is a contradiction on what I have just said. Not only do we must stand back we Also must go back to the Big Bang. This accrued billions of years ago. How many Big Bangs has there been. Your right there has been just the one. We know that it created all the galaxies. Are you with me So far? If this was the case how can the Andromeda galaxy be the oldest? This has got me thinking there must be Something wrong with the astrologer's facts. I have just stated, and I am not the only one with this view point. All the astrologers have deduced this. We have had these galaxies hurtling through space. And there have not been any little bangs. We must understand that the sun's in these galaxies must cool down until the time comes when they explode. This has not happened yet, and the way things are looking we won't see it in our generation. If we take a look how long it is since the Big Bang, then can surmise that our Lord will intervene before this happens. If we look around it looks like our home planet Earth will be the first to go. Why should this be? Our planet is the youngest in the universe. How did I manage to work that out I can hear you say? There was no Big Bang that created Earth. I hope that this has got Some of my readers scratching their heads. This applies to those that do not believe that there must be a creator. Now we will must keep an eye on these non-believers as they will be scratching their heads So much they are going to give themselves an injury. The planets with the smallest sun's are going to be the first that this will happen to. This comes to my observations that being the last galaxy that has been introduced to the universe we have the smallest sun So bang goes us first. Do not start panicking it won't be for a long time yet.

    Sun's hotspots do not have any adverse effect on the Earth's surface. Boffins say that the sun's hot spots, affects its planets.

If you find that any of the below has been repeated, you must look at it as my way of giving you your revision.

72 volcanoes that I how many volcanoes there are in Iceland. There will be more about this subject later.

Japan is the most volatile region in the word for volcanic eruptions. The children when going to school must wear helmets as there is a regular fall of dust and Some of these particles are the size of tennis balls.

I am doing away with chapters as it is a lot of waste space.

Planets going to collide no! no! no!

Bang goes that theory

What black hole.

No ice age

Saturn's rings. The experts don't know. What they do say is that they compose of ice, water and large spaces in the rings.

Human evolution This happened as quickly as the Big Bang but it didn't take as long.

I can light up a black hole. Who's a clever boy then?

Here we go with more theories. The theory they have come up with concerning gas giants is, gas planet is eating up their moons. I have heard of agreed but this is ridicules.

Space And time. What a load of rubbish Some one's watch is on the blink. Tides. If you are relying on the moon to move them, you are going to wait a long time for the tide to go out.

Gravity is not keeping me down.

In trillion of years' time or maybe longer we will not be able to see anything in the universe

The animals were at rest on the ark. As the bible states the lion laid down with the lamb. The lion looked at the lamb with tenderness. He proceeded to lick the lamb.

In the next stall was a laughing hyena. 'That's nice to see,' he said laughingly. The smile was Soon removed from his face when the lion said.

'I am not being friendly I just don't want to forget the taste.'

An eye opener for all space explorers, and for none believers

Not all about physiatrists and religion. Lots of other points made, including my opinion on every day subjects. (Sorry but I need to get my two penny worth in.)

If our moon was created from a large object hitting it where have all the other moons in the universe come from? Have these all come about the same way? If they have not, then we have a problem. Why in the billion of galaxies this has only happened once. I do not know what you are thinking but every time I find a fault with Some of these theories I realise that there must be a creator.

SPACE EXPANDING

You cannot increase nothing. Let us say for augments sake you could. You have nothing and it expands into the space next to it. That space is nothing So how do you know that it has expanded. If you really put your mind to it and try to work, it out. I will send you the prescription the doctor gave me.

No Solar winds on earth. Solar winds cannot escape from the galaxy that it is in.

What do they say about in our stetisphere?

Is there rainbows in other universes galaxies?

The road sweeper was sweeping up after a rock concert. He cleans all around the edge of the park. There is a mountain of rubbish in the centre of the field. The inspector arrives and says to him, how things are going. I have finished he said. The inspector says what about that heap of rubbish in the centre of the field. My mum used to say to me clear the corners and the middle will take care of itself.

They say that all the matter goes through three stages liquid, Solid and gases. I am Sorry I am no chemist but there is Something very wrong in the way these three things are said to go from one to the other. This would mean that there is the same amount of matter in each one. The plasma below the earth's crust is blown out from the volcanoes and earthquakes. When it cools it turns into rock? With the Earth's plates on the move and the rock meets the plasma it turns back into plasma. Any gases that are produced rise into the atmosphere. The cold atmosphere turns the gases into liquid drops by cooling them down. These liquids will Also contain oxygen and nitrogen drops. All these drops then l fall to Earth as rain. These can be seen in a rainbow. The different gases produce different colours. This tells you that the droplets came from different metals. As the contour of the Earth is not on a level plain then the metals in time will finish up on the same level with each other. This can be seen in the iron that is in the Earth's rocks. If the same amount of rocks and plasma was to turn into gases, we would have a fine time dodging them when they decided to fall to Earth. I hope my readers that are Also not familiar with chemistry; science and physics will understand this reasoning.

ASTEROIDS

An asteroid 1 kilometre wide (half a billion tons) was the size that destroyed. Gomorrah the blast covered 300 square miles Earth quake accrues, or an asteroid hit the Earth if you have a look you will see that there is impression left on the surface. As can be seen the Earth has been hit by asteroids of a fair size but we carry on regardless. This must prove the whole of the Earth cannot be destroyed by these occurrences. How about this for a thought? If this asteroid was this big then what size would it take to produce another moon?

You might thing that Some editing has been repeated but you should find there is extra information included.

And after a while God returned to the Garden of Eden and there was nobody about. "Adam, Eve," he called out. There was no answer. "Please show yourself," he cried out, "where you are."

"We are hiding in the bushes," Adam answered.

"Why are you hiding" God wanted to know.

"Because we are naked,"

"You have eaten from the tree of knowledge. I had forbidden you from doing this."

"We have covered our nakedness with fig leaves."

"That is true." Eve said, "I told him there was no need for a fig leaf a privet leaf would have done"

"For disobeying me," God said," woman shall bear children in pain and man's crops will be plagued by pests and weeds."

God forgot to name, Floods, droughts, hurricane winds and Jack Frost. And that was last Monday. Sorry my mistake it was Monday morning

Scientists and all those that deal with the universe have come up that the Sun spins faster at their equator than at their poles. We all know why that is. For all those that would like to be enlightened I will explain later.

They say that an asteroid hit the Earth that caused it to spin.

When the original sun shone there was nothing in the universe to light up. By gum that's a thought. If you think about this, what does it tell you? If you cannot think what it maybe I will light up your grey cells. I will tell you. Although there was nothing there to light the sun's rays they were still travelling outwards. And this kept going out into space until the absolutely zero temperature cooled it down. This had to happen. Otherwise the sun would not have cooled down.

The sun. and black holes. Not as black as it looks. Let's shine Some light on them.

The moon. Is the moons gravity the cause of the tides? Who's pulling whom?

The moons gravity can't pull me never mind our tides.

Ice age. Not as cold as you think.

The earth's spin. Was it knocked for a six?

The Big Bang theories

I can agree on one of their explanations about the Big Bang and that it really happened. They say that the gases in the universe in time will turn into a white dwarf and then into a black hole. I disagree.

D.N.A. species can interbreed but the young are infertile. This basically means that evolution cannot breed a different species. Species do evolve in their own species. This means that no new species can possibly appear.

The land was as one land mass. And it was slowly eroded until there were different continents as can be seen today. This is not passable.

103 days of plague = 4/5 months. This is too long for these plagues to have lasted. What is the most likely time span?

Black holes lit up Dark matter; this has messed up my mind. String theory, this has me all tied up. Black holes. I can see into this.

Also in this book

The Bible v the Charles Darwin's theory and Bang goes the Big Bang theory? Everything you have learned So far will be turned on its head. You think the sun is a powerful force. It's hot but let's cool it down. What has a more powerful force?

The Big Bang. Came in like a lion and will go out like a lamb?

The sun. Not such a bright spark. The colours are a wash out.

The tides. Do you think they are going in and out, think again?

The universe. Is it expanding? Not from where I stand.

Evolution. Are you a man or a mouse?

Space and time. Who's getting younger? My watch is winding down. Everything you have learned So far will be turned on its head. You think the sun is a powerful force. It's hot but let's cool it down. What has a more powerful force?

The national space centre does not know if the universe will go on expanding or it will implode and reverse back to its original state. I have the answers. <u>Find out later?</u>

If the Earth's is spinning on its axis was caused by an object hitting it? What hit the sun's to make them spin? What pull does the moon really have? Ice age means grass eating animals, plants, and insects plus all plants and bacteria would have died out. The answers are scattered throughout the book. The answers to all the questions that scientists, physicists, and biologists have only have theories for? A lot of the following is based on logic there are Some theories. If my logics are scientifically proved wrong, then they are just theories.

Not just for the scientific mind but lots of interest for those who are not inclined that way but would like to know more. There are Also bible revelations. There are So many important points in this book I hope I have put them in the correct order of their importance.

G.M. FOODS

The reason for genetically modified crops, and the reason why we need them. What is the harm that is coursed by the way we grow are crops in today's environment, when this courses untold damaged to our air, water and Soil. The chemicals that are used are ridicules. This could all end tomorrow if all the crops were grown as genetically modified. The arguments against modified crops are given because there is a possibility that new strains of diseases could emerge. Also, there might be strains that Some people are allergic to. The modified crops Also contain viruses and bacteria. There is a possibility. If this is the case, then surly when these products are Sold to the public then there would be a warning on the packet to inform the buyer. There has been a tendency on improving the ability to improve the way we grow our food supply. There is Also the fact that tomatoes and corn has been using genetically modified for the last ten years. I have not heard of there being any new strain of diseases coursed by these crops, if the same procedure was used for other crops then I can see no reason why they cannot be used. The point is Also being made of the damaged that these crops can do to Some insect species. I will give you me reason why these diseases riddled insects are not required for our food supply. (This will open a few eyes.) Looking at the reasons given for not using genetically modified crops seem to be way behind the reason to have them. With all the controversy around genetically modified (GM) foods, Sorting through huge volumes of information can seem like a daunting task. Many members of the public are asking questions about GM foods and they are Also raising concerns about the effects these foods may have on their health or the environment. There are different advantages and disadvantages of GM foods, although to what extent they can help or harm humans and the environment is a debatable aspect of this technology.

BENEFITS OF GM FOOD

The following different bodies that have looked into genetic modified crops have given their opinions. This will save my readers a lot of unnecessary work in the studying the subject. Improved food quality is another benefit asSociated with GM foods. A tomato, for instance, can be engineered to stay fresher for longer, thereby extending its shelf life in the supermarket. Yet another benefit that is believed to occur from GM technology is that crops can be engineered to withstand weather fluctuations and extremes. This means that they can provide sufficient yields and quality despite a severe, poor weather seaSon. Another benefit is that GM foods can be engineered to have a larger content added of vitamins. Also nutrients that are lacking in Some humans can be high content of a specific nutrient that is lacking in the diet can be added. The vitamin A rich 'golden rice' is one example of a GM food that has been engineered to have high levels of a nutrient. As can be seen from the above comments it seems that we must use more modified crops.

ISSUES WITH GM FOODS

Another potential downside to GM technology is that other organisms in the ecosystem could be harmed, which would lead to a lower level of biodiversity. By removing one pest that harms the crop, you could be removing a food Source for an animal. Also GM crops could prove toxic to an organism in the environment, leading to reduced numbers or extinction of that organism. Given that Some GM foods are modified using bacteria and viruses, there is a fear that we will see the emergence of new diseases. The threat to human health is a worriSome aspect of GM technology and one that has received a great deal of debate. Reading a brief fact sheet is a good way to familiarise yourself with the purported benefits and issues related to GM foods. In this way, you can equip yourself with an overview of the knowledge with all of the controversy around genetically modified (GM) foods, Sorting through huge volumes of information can seem like a daunting task. Many members of the public are asking questions about GM foods and they are Also raising concerns about the effects these foods may have on their health or the environment. There are different advantages and disadvantages of GM foods, although to what extent they can help or harm humans and the environment is a debatable aspect of this technology.

BENEFITS OF GM FOODS

GM foods. A tomato, for instance, can be engineered to stay fresher for longer, thereby extending its shelf life in the supermarket. Yet another benefit that is believed to occur from GM technology is that crops can be engineered to withstand weather fluctuations and extremes. This means that they can provide enough yields and quality despite a severe, poor weather seasons. The way global warming is taking effect we are going to need these genetically modified crops Sooner than later. I wonder just how many innovations that the scientists have discovered that has not been of use to mankind. I should imagine that the benefits out way the failures by a long way. I wonder if we take it in another light how many of the failures have done a serious harm to the environment. We know that the spaying of crops to combat the pests have a very bad effect on the environment, but we still use them will we never learn.

They have now found that the plastic beads that are in our toothpastes and in our washing powder to make them more abrasive have entered our sea environmental activist fixated on curbing plastic pollution in the oceans he loves to surf and sail, WilSon focused his efforts on eliminating micro beads, the tiny plastic particles most commonly used as exfoliates in facial scrubs, body washes and toothpastes. Polyethylene plastic beads don't disintegrate or biodegrade, and about 8 trillion beads, — enough to cover more than 300 tennis courts end up in waterways across the United States every day.

REPLY

Owing to global warming and we know what that is doing to the weather all over the world, droughts, hurricane winds, floods. We can all see the damaged that is being done. Can we not see that we will must improve on our abilities to grow more food? It appears that genetic modified crops will must be the way to go. I will explain later the main reason for modified crops and the argument that the pest that will no longer be available for birds as a food Source. I believe that with the reSourcefulness of living things that they would must reSort to a different food supply. Concerning food allergies that Some humans are allergic to there is a warning on the product to tell the consumer that this ingredient is present. That's a clever idea; maybe Someone could come up with that idea on modified food containers. (Tongue in cheek.) Being one that notices things there is one main point to be made and that is concerning mans innovations. It does not matter what improvement we make to help mankind any waste that comes from these products will contaminate the environment and one way or another they will harm Some living organism whether it be animal vegetable or mineral. We are always trying to improve the way we grow our food supply.

I can imagine that Adam was the first one to must try to improve on his food supply. God put a curse on man's crops with pests and weeds. Have you ever thought why with all the vegetation in the world that is growing wild there has not been one epidemic that could destroy this vegetation? If there was ever an epidemic there would be no natural pesticides that nature could produce to destroy these pests. So what on this Earth could possibly produce pests that prefer to eat my vegetables and fruit? What is it that everybody finds a delight to see. They would not harm these beautiful things except me. There a pest and like all pests that was put on this Earth to punish man fore Adams miss doing. Don't get me wrong we are still being punished and that is because of our sins.

I could imagine what Abel would have said if he was an environmental objector. Dad you cannot go about pulling the weeds out of our vegetable crop. You are messing with the environment you have no idea what damage you could be doing. The dandelion weeds keep coming up and the slugs turn their noses up and head straight for my vegetables. What can we see in nature that everybody likes, and I don't? It's the butterfly. They lay eggs that turn into a grub. These eat my vegetables and fruit. They say that they pollinate the flowers. I have seen them on flowers and all they do is spread their wings and sunbathe. The time they spend doing this is ridiculous. All they are doing is preventing the sun from doing any good to my plants. Not only that but while they are there the bees won't go on them. They can live from weeks to a year. This depends on their environment.

We have tried pesticides, but these are not very successive as we must use them on a continuous basic. Come on a cheer for modified crops.

THE BIG BANG

I can agree on one of their explanation about the Big Bang. The universe is expanding. They say that the gases in the universe in time will turn into a white dwarf and then into a black hole. I disagree.

RUBBISH

That Sounds better than I disagree. I think I should study the universe, I mean to say all the studying that the physiatrists and scientists have done and with all their university degrees all they can come up with are theories. Looking at Some of these theories, I think I could come up with theories that would be more believable. But not wishing to be classed in the same vein I will try to use a bit of logic. Don't get me wrong it is by thinking about what they know for a fact that I base most of my logic on. But where I think they are going overboard with Some of their theories I will say So.

Joke

I must tell you a joke I heard. It was on the Colombo T V series. There was this flasher walking down Sun'set Strip coming towards him was a Jewish woman. As they got nearer to each other the flasher opens up his Mac. The Jewish woman looks down at him in disgust and says, call dat... a lining

Giants of the universe Gases are very cold. I had better explain my reasoning for this statement. When a gas is produced it is hot and it rises and So the process starts of cooling it down. Let us take it that gases in the universe must Also be cold then how can they explode into a white dwarf. Here we go again contradictions. This will be explained later as is my wont.

How can the Andromeda and Earth both be the second generation of galaxies? The experts say that the Andromeda is outside the Milky Way and our Solar system is inside the Milky Way. They point out that the Andromeda galaxy is the oldest galaxy and the Earth's Solar system is the youngest. What we need to do is look at this more closely. Put your magnifying glasses away we don't must look that close. I believe that at the time of the Big Bang the Milky Way was formed. As with all explosions the matter that is the heaviest will not necessarily travel the greatest distance. And the matter will not leave the object of the explosion in the same quantity. This then would mean that the largest matter to leave the explosion as in the Andromeda case (the sun) would travel further. You might ask how I have come to that conclusion. We're talking about the first sun in our universe. Do you think that all the sun's produced by the Big Bang would be any different? No of course not that means we shall look at the last sun that appeared in our universe. That's our sun. We shall look a bit closer, not to close we don't want anyone getting burned. What do we see? I will jog your memory; there are hot spots these shoot out at a tremendous speed. This tells us that when the original sun obtained a crust that covered nearly the entire surface and it became like a pressure cooker until the pressure became that great that it exploded. This then tells you that Some of the matter travelled further whether it is large or small. What does this tell you? All the sun's in each galaxy did not have the ability to hold this matter into its orbit. Otherwise it would have been held in a closer orbit to it. Another way of looking at it is as this matter travelled further away the pull of the sun's penny force becomes stronger. I will explain later my deductions on the working of the sun's abilities. All though gases are different they would be spread out and come together later. Our Earth is quite another story. Dark matter. They are saying that dark matter is drawing the galaxies together.

Brett Williams

WRONG

I will explain later why I have come to this conclusion.
How old is the universe?
How old is the Universe? Until recently, astronomers estimated that the Big Bang occurred between 12 and 15 billion years ago. Our Solar System is thought to be 4.5 billion years old and humans have existed as a genus for only a few million years. Scientists believe that the Solar system was formed when a cloud of gas and dust in space was disturbed, maybe by the explosion of a nearby star (called a supernova). This explosion made waves, in which squeezed the cloud of gas and dust together. Wrong (I am glad I was nowhere near them at that time.) This cannot have happened. I will explain why later.

How did the terrestrial planet form?
Formation of Jovian planets: In the outer Solar nebula, planet similes formed from ice flakes in addition to rocky and metal flakes. Since ices were more abundant the planetesimals could grow too a much larger size, becoming the cores of the four Jovian (Jupiter, Saturn, Uranus, and Neptune) planets.

And just where did the ice come from? The contents of space between the heavenly bodies are zero. If as they say in the above that there is ice in space then all of space would contain ice. All the matter from the Big Bang finished up in one of three components. There are the gases, molten plasma containing metals and rocks Also containing metals. The ice would only appear when conditions of the molten plasma objects gave off an atmosphere that contained water. This can only happen if there is water on that planet in the first place. It would not enter space because the spinning of that planet sun would hold it its gravitation field. So you can see my problem with space containing ice. If space did contain ice then there would be a lot more of these terrestrial planets.

The Andromeda Galaxy known as Messier 31 or M31 is a spiral galaxy located about 2.5 million light-years (2.4×10^{19} km) from Earth. Located in the Andromeda constellation, it is the closest spiral galaxy to the Milky Way. Where as our Solar system is a long way from the Milky Way

REPLY

How did the scientists work out how far away the Andromeda galaxy is in compariSon too Earth's distance. They are both *supposed* to be the second generation of galaxies formed. Why then is there such a massive difference. The Milky Way's galaxies were all formed at the same time. This to me means that they are all the same age. So why is the age of Andromeda age of 9,006? The age of the earth is 4.543 billion years. The universe is 15 billion years old. They have been wrong about numerous statements concerning the age of planets. Most of these statements are based on the amount of radiation decay that the rocks contain. This brings me back to the original sun. It has all the elements that are now in the universe, this includes the Earth. Bang and here we are. Not right away as my belief in a creator means that the Earth was created after the Big Bang. But the same scenario still applies. The Earth was created from fire and rocky matter and not long after our creator gave us a moon. This was to light up our night sky. Why would a creator take So long before he decided to produce another planet? Time has no meaning for a God that has been in space for trillion of years. The universe is said to be fifteen the billion years old. To me that mean's that the plasma that is here on Earth is the same age as the original sun's contents. Never mind about the effects of radiations decay it does not change the age of the plasma. I looked up on the internet about the effects of radiation and the metals decaying and producing a different metal. If you have two or more metals in a liquid state, they do not make another metal. The heaviest metals as with gases sink to the bottom. We have all these different metals in the Earth's crust and they are all separated. Originally, they were all molten plasma. How's this for Something that the scientist, geologists can get their teeth into, what age does the plasma show when it has cooled down to the state where it can be tested?

FACTS ABOUT THE ANDROMEDA

The Andromeda Galaxy gets its name from the area in which it appears, the constellation of Andromeda – which is named for the after the mythological Greek princess Andromeda. Andromeda is probably the most massive galaxy in the Local Group. This is contrary to previous research which speculated that the Milky Way was the most massive because it contained more dark matter. A 2006 study put the Milky Way at around 80% the mass of the Andromeda Galaxy. There are around 1 trillion stars in the galaxy, compared with the Milky Way which has around 200-400 billion. In approximately 3.75 billion years, the Andromeda and Milky Way galaxies will collide, merging to form a giant elliptical galaxy. A team of astronomers in 2010 believe that M31 was formed between 5 and 9 billion years ago when two smaller galaxies collided and merged.

WRONG.

If the experts know anything about the universe they should know that the way the universe works makes this impossible.

REPLY

We are always being told that the universe is expanding. And then on the other hand there will be a time when the universe will collapse inwards. As a beginner of the universe workings how can we be expected to know which is right. I therefore will go back to the beginning. This is of course The Big Bang. This is a known fact that the universe is expanding. Doing a bit of elementary maths then the universe is not going to collapse into itself. If it was it would have happened not long after the Big Bang. As everything after the Big Bang was reduced in size. This means not only the matter but the power of each sun. Now let us look back. What are they saying that coursed Andromeda to become the size it is'. It was coursed by two smaller sized galaxies colliding. Wrong. We know this cannot happen. Let us now look at the Milky Way this has the brightest light. The scientists say it is from gases and they will get that hot they will explode into a white dwarf/Nova. Gases do not explode from a heat they must be ignited by a spark. If there was electrical universe then any gases would be ignited before it got to the stage when it would explode to such a degree. When a gas explodes the amount of Solid matter that it produces is So minimum that it is not even worth considering. Let us take my diagnosis. The reason that the Milky Way's light is the brightest in the universe it is 98% of the sun that remained after the Big Bang. It is the same as our Solar system. Our sun contains 98% of the matter in our Solar system. This goes to prove that there is such a time lag between the two, but the result is the same. Now why would the Andromeda galaxy have more stars than the Milky Way? My explanation is but don't take this as gospel the original sun cooled down, as our planet is doing but you can see what is happening with our planet. We have volcanoes and earthquakes. Let us see what would happen when the time comes that the earth was covered by a crust. The pressure would build up until it exploded. There would not be an even eruption there would be weaker crusts this would mean that the matter in these places would be blasted out into space. This matter, being made up of Solids, molten plasma and gases. The molten lave would of course be the star and this would have the power to form a galaxy. The matter that is left would of course be in the Milky Way. That is why it is has the largest sun. The Andromeda Galaxy has an apparent magnitude of 3.4, which makes it bright enough to be seen by the naked eye on moonless nights. The Andromeda Galaxy is approaching the Milky Way at rate of around 110 kilometres per second (68 mi/s. Let me explain to the physitrists, scientists and all the other people that give us information on the universe. If the universe is expanding how is this possible? I will must explain to the scientists, physitrists and the boffins how the universe works. I am not trying to be clever I am going by what these boffins have been saying. Suddenly, the way the universe works has been turned on its head. Galaxies cannot be pulled together and if they say it is gravity, you will get my reason why this is not possible. The age of the Andromeda galaxies is 9.006 billion years. Can two galaxies move away from each other faster than the speed of light? Yes, say the powers to be. My answer.

RUBBISH.

The Hubble Constant is the measure of how fast the Universe is expanding today and its value has been measured to be 70 km/s per Megaparsec (a parsec is just a unit of distance equal to about 3.26 light-years, and a Megaparsec is a million parsecs). This means that on average, for every Megaparsec two galaxies are separated by, they are moving away from each other by 70 km/s. Therefore, to be moving away from each other at the speed of light, two galaxies would need to be separated by a distance of about 4,300 million parsecs. This is smaller than the radius of the observable Universe, therefore not only are there galaxies in the Universe that are moving away from us faster than light, but we can still see them! This raises two additional questions: If another galaxy is moving away from us faster than light, there would come a time when we would not see it. Isn't it a violation of the theory of relativity to have two things moving apart faster than the speed of light? I will explain later why the above maths has it wrong. Light is still the fastest thing in the universe this is the Einstein's calculations of how the universe works. I am Sorry I do not agree. The answer to the first of these questions is that the light the distant galaxy is emitting today will never reach us, So we will never know what it looks like today. This is because today it is moving away from us faster than light, So the light it emits doesn't travel fast enough to ever reach us. However, the light that it emitted billions of years ago, when the Universe was smaller (remember it has been expanding all along) and when that galaxy wasn't receding from the Milky Way as fast, is what we are seeing today. In other words, we are seeing that galaxy as it was billions of years ago. The second question is an interesting one that confuses many people. The theory of relativity does indeed state that nothing can travel faster than light, however this refers to motion in the traditional sense, meaning you can't launch a spaceship and travel through space faster than light. The two galaxies we've been discussing are not travelling through space; it is the space between them that is expanding. Rubbish. Or put in another way, they are stationary and all the space around them is being stretched out. This is why it doesn't violate the theory of relativity, because it is not motion in the traditional sense. The explanation that the expanding galaxies are in fact travelling at a speed that is faster than light. RUBBISH. Part of question one is that as time passes we will not be able to see these stars as they are travelling faster than light. At what time would these stars become impossible too see? If there is nothing in space to reduce the strength of a sun's heat or light, then it cannot reduce its power. We only must look at our planet Earth and we can see the difference on a cloudy day but when the clouds clear the sun's rays are back to what it was before. I will explain later why the two galaxies rushing apart do not increase the speed So that is travels faster than light. As time passes by and the Big Bang was the first stage of our universe. This created the galaxies in the Milky Way. Each galaxy is controlled by its sun. The original sun in the sky cooled down to the extent that it exploded. This is what coursed the Big Bang and that was the start of our universe. This sun was of a colossal size that when it exploded it produced all the sun's/stars, asteroids, planets and gases. We must remember about forces balancing things out. Only a small amount of the sun was blasted into space and the rest was not affected. The bright light that can be seen in the Milky Way is not a super nova's it is the original sun. This will happen with the second generation of bangs. There is a rumour going around that the second generation of galaxies consist of the Andromeda galaxy and our own Solar system. The reason they say that is because the Andromeda galaxy is So far away from the Milky Way. Sunlight is controlled by its galaxy as is the gravitational field, and each galaxies field is controlled by other galaxies around it. The universe is the only force that is different as these controls all the galaxies. It's strange but you might have noticed that there has been a lot of talk about the second generation of these are the same of ones but older. And each galaxy is expanding. The size of these galaxies depends on the size of the sun. The bigger their sun's are then the bigger their galaxies. Here is were we have Isaac Newton's calculations on equal forces balancing things out as gravity. The force from each galaxy when they come together they will balance themselves out as the stronger galaxy encroaches on the weaker force there will come a time when the smaller galaxies force will be strong enough to halt the other force. This happens at the blink of an eyelid. It's probably faster, as a Nano second. Here it is, these two might be rushing apart but they are controlled by the universes sphere movement. Being a lithan gas the sun's rays do not get affected. Why is this, a gas is one of the lightest elements the only thing that I can see is that it is travelling So fast. You might ask how it is possible that we have gases around the sun's and we Also have gases at the far edge of these galaxies. When a sun decides it is time it spread its wing and investigated the future. The sun gases that where around its circumference remained there. The sun's that went into space left a trail of gases in its wake, as a comet leaves a trail of condensation in its wake. When it finally comes to rest its gases then are around its circumference. The gases in space are swept together at the galaxies weakest points.

FASTER THAN LIGHT?

The explanation for this is because if two galaxies are travelling away from each and each one has the light of the sun that is travelling at the speed of light then that increases their speed by two. I have a fault with this mathematical sum. But do these boffins not know how the galaxies work and how the maths used in this scenario does not compute. One reason is the galaxies are travelling at seventeen thousand miles a second and each galaxy is travelling at this speed and they are all travelling away from each other. I will tell you how I have come up with the mathematical statistics out You might as well ask how is it possible that Someone who didn't have the basic secondary education needs to explain to the scientists, mathematicians, physitrists, and this includes those that are working for N.A.S.A. how is this possible. Now we have the internet I can at the touch of a button see what the boffins are saying. I have noticed that the points they make keep changing. If these are supposed to be our top brains in the country, why do they not come up with the same answers? I then decided to use my own deductions on the information that they were giving. It is there for all to see So I will now explain how I have come to the decision that light is still faster than the expansion of the universe. Let us look at the way different matter/rays are affected by the working of the universe. First the sun's rays are not affected by gravity. So this means its speed is consistent. This is 299,792,458, metres per second. There is another way to decide if in fact the speed of light is the fastest thing in the universe. With the technology we have now in space exploration and the things we are learning why nobody has thought of using this in finding out the speed of light. If there is anything faster, they have not found it. I have.

If we use Isaac Newton's discovery that force has an equal from one direction has an equal force from the other direction, then it counter reacts. Now let us look if this Also applied if it occurred in space. If this was So then when the Big Bang occurred 2% of the mass of the Sun was blasted into space and the universe was created. That means that in the billion of galaxies there must be billions of sun's. When the time comes for their demise and they explode what do you think will happen? Let us look at our home planet. We already have planets, moons, gas planets and asteroids. Will these be blasted to high heaven? Maybe they will crumble into smaller objects. We will take at face value what the astronomers say. The universe is travelling at fifteen thousand miles per second. They will have no affected on other objects because they are travelling at the same speed but are millions of years apart. What will change is the size of these sun's, planets and asteroids. That means that the rest of the objects are only five percent of that sun's size. This is after the Big Bang. This must be the brightest star in the sky. Looking at photos of the universe I would say it was at the centre of the Milky Way.

The sun's light and heat has been decreasing since the time it formed. This occurs through two processes in nearly equal amounts. First, in the Sun's core, hydrogen is converted into helium through Nuclear fusion, in particular the p–p chain, and this reaction converts Some mass into energy in the form of gamma ray photons. Most of this energy eventually radiates away from the Sun. Second, high-energy protons and electrons in the atmosphere of the Sun are ejected directly into outer space as a Solar wind. The original mass of the Sun at the time it reached the main sequence remains uncertain. The early Sun had much higher mass-loss rates than at present, and it may have lost anywhere from 1–7% of its natural mass over the course of its main-sequence lifetime The Sun gains a very small amount of mass through the impact of asteroids and comets. (This cannot be right as it would burn up long before it reached the sun's surface.) However, as the Sun already contains 99.86% of the Solar System's total mass, these impacts cannot offset the mass lost by radiation and ejection.

As we can see from, https://en.wikipedia.org/wiki/Solar mass. In the above information that in our Solar system the sun contains 99.86% of the mass of our Solar system. 'What does this tell us?' it's as plain as the moon at night. What happens when each galaxy is formed from the first Big Bang? You have exactly what you have in our galaxy/Solar system. Gases, plasma from our sun, and Solids in the earth's crust. You might hear the experts say that the galaxies are changing in the way they came about. They do not. To get back to my question. As can be seen, that two% of the sun's matter is released when the sun explodes. This happens with every galaxy, I mean to say if scientists do an experiment and each one turns out with the same result. We then take that as concrete evidence. This then proves Isaac Newton's theories as facts. The outer edges of the sun's crust which are mostly Solids. There would Also be plasma blasting out into space. Now what we had was a pressure cooker, the sun that was getting hotter and hotter, under the crust. As this crust got thicker and thicker the heat got to the point that there was only one way to go was outward. You now have pressure coming from all sides this is where Isaac Newton's force comes into play. With this equal force it all comes to a stand still. This shows that all the sun's in each galaxy are the remains of a bang and the planets, asteroids and the gases came from that bang. Why do the boffins say that the amount of stars in a cluster is So many?

The distance of a cluster

Reply. Why this is when you see a photo of a stars cluster the centre seems to have that many more star's and they are closer together. This is because of the angle that they are in compariSon to where we are. As the universe expands and the galaxies that were created after the Big Bang you would think that there would be more space between the galaxies. They will appear to be closer because they are further away. This is what you see if you look down a railway track the rails seem to finish up at the same point. This not recommended as you never know if a train is due as they are never on time. Let us take it a bit further or use the train scenario as if it was the universe. We have not one set of tracks but for arguments sake we have a dozen tracks. The end of these tracks would finish up at one point. The point of these tracks would appear as one small spot. With a star constellation they would not finish up at one point as the light from these galaxies come under my penny forces. This is my way of showing my logic of how forces react with the same force. More about this later

> Why do they keep finding new galaxies and new planets in our Solar systems? This is because of their position in the universe. Mainly because we orbit our sun at irregular distances and So we see the universe at a different angle.

The ocean covers more than 70 percent of the surface of our planet. It's hard to imagine, but about 97 percent of the Earth's water can be found in our oceans. Of the tiny percentage that's not in the ocean, about two percent is frozen up in glaciers and ice caps. Less than one percent of all the water on Earth is fresh. A tiny fraction of water exists as water vapour in our atmosphere. According to the U.S. Geological Survey, there are over 332,519,000 cubic miles of water on the planet. A cubic mile is the volume of a cube measuring one mile on each side. Of this vast volume of water, NOAA's National Geophysical Data Centre estimates that 321,003,271 cubic miles is in the ocean. That's enough water to fill about 352,670,000,000,000,000,000 gallon-sized milk containers! I am not certain about this amount in fact to be more accurate I would give or take a few. I can understand that Some of my readers might be baffled by the above statistics and the following statements. They are not the only ones. There is a perfectly good reason for this. As you read on I hope it will become apparent. When I divulge my results, you will be amazed and Also get a sinking feeling. This will not only make you step back and say, I am amazed, it will Also make the scientists, physitrists and geologists say, I never thought of that. There is another section of Society that will be delighted that they are proven right all along. I will not reveal which sector this is. Maybe with a bit of thought you might come up with the answer. I like to keep my readers on their toes. If the truth be known. I do not know how I managed to work this out. Let us look at the following first statement. You will see the water covers a lot of the Earth's surface. I would presume that my readers who do not have a lot of knowledge would have heard of that fact. If they haven't then they know now. What this tells you is So obvious that I cannot understand why the scientists, physitrists and geologists have not realised this. I am Sorry to leave you sitting on the edge of your seat but it is too important to disclose at this date.

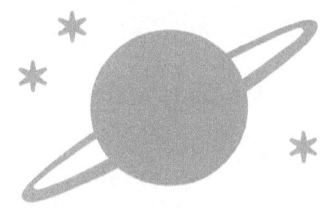

STRUCTURE OF THE EARTH

From Wikipedia, the free encyclopaedia. The most northern point of the earth is the thinnest part of the earth's crust. I take it that this must be three miles thick as the geologists say it varies from three miles thick to forty-four miles thick. What this tells me must do with the amount of water that the earth has in its oceans. It Also recalls the scientists and the geologists working out the contours of the land mass and how over time they have eroded in such a way that if pushed together they would fit together like a jigsaw. I have a problem with this. Okay I can hear you say I am no good at jigsaws. The interior structure of the Earth is layered in spherical shells, like an onion. These layers can be defined by their chemical and their rheological properties. Earth has an outer silicate Solid crust, a highly viscous mantle, a liquid outer core that is much less viscous than the mantle, and a Solid inner core. Scientific understanding of the internal structure of the Earth is based on observations of topography and bathymetry, observations of rock in outcrop, samples brought to the surface from greater depths by volcanoes or volcanic activity, analysis of the seismic waves that pass through the Earth, measurements of the gravitational and magnetic fields of the Earth, and experiments with crystalline Solids at pressures and temperatures characteristic of the Earth's deep interior. You could say that I am putting the cart before the horse but there is no horse. You might think I have got this the wrong way around. But the answer is to a question that has not been asked. Why are the gas planets further away from the sun, and the rocky planets nearer? I worked it out by myself about the rocky planets being nearer. For beginners that want to look into space and learn a bit about gas giants. Gas giants describe a planet that is not composed of mostly rock and other Solid substances. Gas giants are almost entirely formed of various gases. These planets are not completely gas though. At the centre, this is what astronomers call a rocky centre. This term is Somewhat misleading though because the rocky centre is liquid compounds, including molten heavy metals. The term was created by James Blish, a science fiction writer from the mid-1900.

Gas giants are Also called Jovian planets after Jupiter, the prototype of gas giants in our Solar System. There are four gas giants in our Solar System – Jupiter, Saturn, Uranus, and Neptune.

The gas giants in our Solar Systems have several similar characteristics. All our Solar System's gas giants are outer planets, which means they are the furthest planets from the Sun. Compared to terrestrial planets, gas giants are extremely large and massive. For example, Jupiter has a mass 318 times the mass of Earth, which is a terrestrial planet. Despite their size, gas giants are low-density planets because they are composed almost entirely of gas. In addition to being large, these planets rotate extremely quickly. Jupiter rotates So quickly that it has flattened at its poles. The gas giants are extremely cold planets, although that is mostly since they are far from the Sun. Gas giants Also have dozens of satellites and ring systems. Saturn is famous for its beautiful rings, which can be seen with the unaided eye from Earth. Mystery of how Saturn's rings formed is Solved: Rock and ice were repeatedly smashed together to create planetary wonders.

WRONG

Saturn's rings formed in circles that are dominated by one particle size Most common particles are smaller than a metre while larger ones are rare.

Scientists have calculated the rings appear to have reached a steady state due to repeated aggregation and breaking apart of the particles with in them. They say this Also explains how rings around other planets have formed. Why, if this has been happening over billions of years who has put the brake on?

REPLY

If the above is true, then all the asteroids by this time would most certainly be dust particles. If there is anybody contradicting this then, think. If all of them were not disintegrated there certainly would be a larger number of larger ones.

By Richard Gray for Mail Online

They have captivated and mystified astronomers since they were discovered 400 years ago, but scientists may have Solved the mystery of why Saturn's rings look like they do. Mathematicians have analysed the neat distribution of particles of ice and rock throughout the rings and say they are characteristic of catastrophic collisions over time. Their findings help explain why there is such an abundance of small particles in the rings, while objects wider 32-ft (10 metres) are incredibly rare. Saturn's rings, shown in the image above, are made up of particles of varying size. The different colours in the image above represent different particle size with green being the smallest. Researchers have found the distribution of ice and rock through Saturn's rings is a result of how the particles have collided and separated.

Recent research has suggested that each ring layer is dominated by a particular particle size and formed relatively recently in the planet's history.

The research may Also help to predict the presence of rings around other planets.

How Saturn is devouring its moon: Stunning images revealed. Scientists believe that the large particles appear to have collided at slow speeds, and in turn triggered smaller particles to collide at higher speeds.

Stunning images sent back by the Cassini spacecraft has revealed how Saturn's icy moon, Enceladus, is slowly being eaten up by the gas giant's rings.

The 310 mile (500km) wide moon, which astronomers say may harbour alien life, has a network of geysers pumping out tiny chunks of ice-water at 800mph (1287km/h). (I hope these aliens have waterproof clothing.)

But these geysers won't be there forever as images show how long, sinuous, tendril-like structures near Enceladus are transferring material from the moon into Saturn's rings.

These ghostly tendrils have long been known to follow Enceladus in its orbit around the gas giant – but the images provided scientists the first opportunity to track their Source.

The tendrils reach into Saturn's E ring - the ring in which Enceladus orbits - extending tens of thousands of miles away from the moon.

This results in a cycle of aggregation and fragmentation until a steady state is reached.

The researchers say this can Also explain the formation of rings around planets like Uranus, Jupiter and Neptune, along with distant asteroids Chiron and Chariklo.

It could Also help scientists to predict how the rings around planets outside of our own Solar system may look.

ProfesSor Nikolai Brilliantov, from the University of Leicester who led the study, said: 'Saturn's rings are relatively well studied and it is known that they consist of ice particles ranging in size from inches to about 32 ft (10 metres).

'With a high probability these particles are remains of Some catastrophic event in a far past, and it is not surprising that there exists debris of all sizes, varying from very small to very large ones.

'We have finally reSolved the riddle of particle size distribution. Our study shows that the observed distribution is not peculiar for Saturn's rings, but has a universal character.

'In other words, it is generic for all planetary rings which have particles to have a similar nature.'

ProfesSor Brilliantov and his colleagues examined the size distribution of ice and rock material in the rings of Saturn to develop a model that may explain their formation.

The explanation given above does not give a true picture of how the universe works. If all the objects in space were colliding then over billion of years I would imagine that there would not be many large objects left. The way the sun's control the galaxies shows that it is not possible for objects in space to keep on colliding. Do the boffins now know how the universe works? I will enlighten them later.

From Wikipedia, the free encyclopedia

In the Solar System, ice is abundant and occurs naturally from as close to the Sun as Mercury to as far away as the Oort cloud objects. Beyond the Solar System, it occurs as interstellar ice. It is abundant on Earth's surface – particularly in the polar regions and above the snow line and, as a common form of precipitation and deposition, plays a key role in Earth's water cycle and climate. It falls as snowflakes and hail or occurs as frost, icicles or. Ice molecules can exhibit seventeen or more different phases (packing geometries) that depend on temperature and pressure. When water is cooled rapidly (quenching), up to three different types of can form depending on the history of its pressure and temperature. When cooled slowly correlated proton tunnelling occurs below 20 K giving rise to macroscopic quantum phenomena. Virtually all the ice on Earth's surface and in its atmosphere is of a hexagonal crystalline structure denoted as ice Ih (spoken as "ice one h") with minute traces of cubic ice denoted as ice Ic. The most common phase transition to ice Ih occurs when liquid water is cooled below 0°C (273.15K, 32°F) at standard atmospheric pressure. It may Also be deposited directly by water vapour, as happens in the formation of frost. The transition from ice to water is melting and from ice directly to water vapour is sublimation.

I do not know if the above information is of any value to anyone. Space in its own right does not contain any water if it did the sun's rays would be able to illuminate it as in turning it in and with the temperature being at absolute zero ice would have formed, and this would Also show up.

This means that there is no point for anyone interesting in ice or requiring ice to look to the high heavens. There is more in your freezer or wait until it snows. As can be seen there are Some lovely designs. These can be used in many ways. There's crocheting for males that way inclined. Curtains are another way that they can be used. You will must get the wife or girl friend to hang them because woman say men cannot hang curtains. Here's a thought. What is the minimum amount of people does it take to change a light bulb. The answer is two. Females won't change a light bulb it takes a man, and a woman holding the spare light bulb and telling him what to do.

They say that the gas planets are very cold. They Also have the theory that there are places in space that consist of nothing but gases and in time they will explode into a super nova So creating another star. These gases need to get organised and decide which group they belong to. In my experience of gases when they explode they do not create a Solid what they do create is another gas. If there are any Solids from these gases exploding the Solid content would be very small. There is a Solid made possible by two gases. The answer will be at the end of this paragraph. You might ask how I came to this decision concerning salt. The main reason given by experts is that it was carried there by rainwater washing it from the rocks. As salt is disSoluble and rocks and stone do not easily disintegrate as quick under these circumstances they remain. If this is the case, then the rocks are inclined to retain water and when the water freezes then they would disintegrate with these weather conditions. That's my problem with their scenario. You might as well ask okay clever clogs what's your explanation. Let us keep to the scenario of it coming from the land. This would mean that the beeches would have as much salt as sand. And most of the lands coast line would be saltier than the sea. So now you see my problem, there are two elements when mixed that turn into salt and they are Sodium and chloride. These two gases when mixed turn into salt. These do not occur on the land. So doing a quick calculation I have come up with a decision they came from the sea. The Sodium comes from the rocks and mixes with the chloride. This comes from gases produced from volcanic activity. Getting back to the gas problem, do not misunderstand me it has nothing to do with my digestion. Astronomers have Also discovered gas giants around stars in other Solar systems. In fact, these are the only extra-Solar planets that scientists have been able to discover yet. These extra-Solar gas giants seem like Jupiter and the other gas giants in our own Solar System. Astronomers have been studying these planets using powerful telescopes, but they have not been able to find out much information about them So far. Some astronomers are searching for life on these planets. They have discovered Some extra-Solar planets in the habitable zones of other Solar systems, and they believe that life could exist on these extra-Solar planets or at least the moons of these planets. How the hell you expect us beginners to believe that anyone can stand on this type of surface I do not know. If people are going to land on these gas giants I would imagine there would be a weight restriction.

Because the gas giants are farther away from Earth than the terrestrial planets, astronomers have not been able to study the gas giants extensively up close. Hopefully, that will change as NASA sends more spacecraft out to explore the outer planets. The two paragraphs below are for the space boffins.

If you are looking for more information on gas giants, look at NASA's planets and think. Quest's habitable moon's around extra-Solar gas giants. On what basic do they presume this?

Universe today, has a number of articles on the gas giants including gas giants gobbled up most of their moons and the Jovian planets.

REPLY
RUBBISH

To save you the problems of reading pages about this information I have wrote about their basic theories and then give my logical answer. If these gas giants gobble up their moons and Jovian planets how can this be possible. So this is a gas gobbling up gases, at this rate trying to figure this out has used up more of my grey cells that I can afford to lose. The only way that this could happen is by the Boffins explanation that it is by the pull of gravity. If gas giants gobble up moons there would be nothing left in space. We have all the gases here on Earth that finish up in our atmosphere and not one atom of Earth has been devoured.

My answer to the above statement I hope this will have you thinking. By Jove he's got it, I think he's got it. By my reckoning it is a better explanation than the above scenario. For all those that are lost join the queue. When I am giving my reasons why the magnetic fields change, I suppose as they say a change is as good as a rest. I hope that my readers that have just a touch of this subject will see that my explanation can be seen as reasonable. The scientific explanation below says that the magnetic fields wander causing the north to move to the South. They do not say what causes this. Surly they can come up with a theory. I have the explanation. I will explain all the above at a better time.

The following article gives many theories about black holes. I do not know if it possible to put holes into a black hole theory but I will explain why they have got it wrong. They have Also come up with theory/deductions concerning the tides movement. I will put this to rights. They keep coming up with these theories and I will keep knocking them down. I will Also put a hold on their gravity theories. I love this job. As I have said I will reveal the answers to the above at a more appropriate time.

I have spent a lot of time on the following subjects. (In fact, there were thirteen pages and that accounted for months of researching) These are the magnetic forces around earth and those radiating from our sun. Also the Solar winds and how they affect Earth and our moon. There were photos showing how these things work. I have deleted all these pages and photos as I did not see the point of boring my readers with all these theories. I could not see the point as I can give my point of view about them in couple of jiffs. What they where saying about Earth's magnetic force that came from our poles and went into space. I had to think about this for a while, and I came to the conclusion that if the sun's rays can get through our ozone layer then I could not see why magnetic wave could not radiate outwards. Plus, if they couldn't then our atmosphere would have that much magnetic interference that all our communications would have that much unintelligent crackling we would not be able to separate from a politician's debate.

Another one of their theories concerned the moon controlling our tides. I will let you into a secret So keep it to yourself. Not now later.

There is a paper below on our moons ability to control are tides. They Also state that the sun has the same affect.

Tidal triggering of earthquakes
From Wikipedia, the free encyclopedia

Need to find more Amplitude of the ocean tide at <u>Golden Gate Bridge</u> for five weeks in 1970.

Tidal triggering of earthquakes is the idea that tidal forces may induce seismicity.

In connection with earthquakes, syzygy refers to the idea that the combined tidal effects of the sun and moon either directly as earth tides in the crust itself, or indirectly by hydrostatic loading due to ocean tides should be able to trigger earthquakes in rock that is already stressed to the point of fracturing, and therefore more earthquakes should occur at times of maximal tidal stress, such as at the new and full moons. Indeed, recent work has concluded that "large earthquakes are more probable during periods of high tidal stress.

Previously, scientists have searched for such a correlation for over a century, but with the exception of volcanic areas (including mid-ocean spreading ridges) the results have been mixed. It has been suggested that Some negative results are due to failure to account for tidal phase and fault orientation (dip), while "many studies reporting positive correlations suffer from a lack of statistical rigor. One systematic investigation found "no evidence for an increase in seismicity during intervals of large tidal range but there is clear evidence for small but significant increase in earthquake rates near low tide"; it did not find an increase of earthquakes near peak spring tides. Seismicity is favoured at low tides, particularly for reverse faults, because unloading unclamps the fault, reducing friction. Ocean loading has no effect at all on strike-slip faults.

Research work has shown a robust correlation between small tidally induced forces and non-volcanic tremor activity. Volcanologists use the regular, predictable Earth tide movements to calibrate and test sensitive volcano deformation monitoring instruments. The tides may Also trigger volcanic events.

I have written why gravity is the strongest power in the universe.

Saturn's rings are inside the orbits of its principal moons. Tidal forces oppose gravitational coalescence of the material in the rings to form moons.

In the case of an infinitesimally small elastic sphere, the effect of a tidal force is to distort the shape of the body without any change in volume. The sphere becomes an ellipSoid with two bulges, pointing towards and away from the other body. Larger objects distort into a void, and are slightly compressed, which is what happens to the Earth›s oceans under the action of the Moon. The Earth and Moon rotate about their common centre of mass or barycenter, and their gravitational attraction provides the centripetal force necessary to maintain this motion. To an observer on the Earth, very close to this barycenter, the situation is one of the Earth's as body 1 acted upon by the gravity of the Moon as body 2. All parts of the Earth are subject to the Moon's gravitational forces, causing the water in the oceans to redistribute, forming bulges on the sides near the Moon and far from the Moon.

When a body rotates while subject to tidal forces, internal friction results in the gradual dissipation of its rotational kinetic energy as heat. In the case for the Earth, and Earth's Moon, the loss of rotational kinetic energy results in a gain of about 2 milliseconds per century. If the body is close enough to its primary, this can result in a rotation which is tidally locked to the orbital motion, as in the case of the Earth's moon. Tidal heating produces dramatic volcanic effects on Jupiter's moon Io. Caused by tidal forces, Also because a regular monthly pattern of moonquakes on Earth's Moon. Tidal forces contribute to ocean currents, which moderate global temperatures by transporting heat energy toward the poles. It has been suggested that in addition to other factors, harmonic beat variations in tidal forcing may contribute to climate changes. However, no strong link has been found to date. Tidal effects become particularly pronounced near small bodies of high mass, such as neutron stars or black holes, where they are responsible for the "spaghettification" of infalling matter. Tidal forces create the oceanic tide of Earth's oceans, where the attracting bodies are the Moon and, to a lesser extent, the Sun. Tidal forces are Also responsible for tidal locking, tidal acceleration, and tidal heating. Tides may Also induce seismicity.

By generating conducting fluids within the interior of the Earth, tidal forces Also affect the Earth's magnetic field.

MY OBSERVATION

Well I am glad they Sorted that out. I hope you found the above interesting most of it is going over my head. This is beside the point I am going to put forward is my reason why the moon and the sun's gravitational pull has no effect on the tides here on Earth. This should raise a few eye brows. As anything that occurs in our galaxy will Also have the same effect in the universe

Earth's magnetic field, Also known as the geomagnetic field, is the magnetic field that extends from the Earth's interior out into space, where it meets the Solar wind, a stream of charged particles emanating from the Sun. Its magnitude at the Earth's surface ranges from 25 to 65 Microtel's (0.25 to 0.65 gauss). Roughly speaking it is the field of a magnetic dipole currently tilted at an angle of about 10^0

with respect to Earth's rotational axis, as if there were a bar magnet placed at that angle at the centre of the Earth. The North geomagnetic pole, located near Greenland in the northern hemisphere, is actually the South pole of the Earth's magnetic field, and the South geomagnetic pole is the North Pole. Unlike a bar magnet, Earth's magnetic field changes over time because it is generated by a geodynamic (in Earth's case, the motion of molten iron alloys in its outer core). While the North and South magnetic poles are usually located near the geographic poles, they can wander widely over geological time scales, but sufficiently slowly for ordinary compasses to remain useful for navigation. However, at irregular intervals averaging several hundred thousand years, the Earth's field reverses and the North and South Magnetic Poles relatively abruptly switch places. These reversals of the geomagnetic poles leave a record in rocks that are of value to paleomagnetists in calculating geomagnetic fields in the past. Such information in turn is helpful in studying the motions of continents and ocean floors in the process of plate tectonics. The magnetosphere is the region above the ionosphere that is in space. It extends several tens of thousands of kilometres into space, protecting the Earth from the charged particles of the Solar wind and cosmic rays that would otherwise strip away the upper atmosphere, including the ozone layer that protects the Earth from harmful ultraviolet radiation The following concerns Solar wins and the magnetic fields. The notes below are only a small portion of what has been written by quite a number of scientists; with information they have received from a number of space ships that have been sent to explore our Solar system. There have been probes by the Americans and the Russians. I did not want to recall what has been said as there would be enough to write a book about.

Related phenomena include the aurora (northern and Southern lights), the plasma tails of comets that always point away from the Sun, and geomagnetic storms that can change the direction of magnetic field lines. As comets are basically ice blocks the heat of the sun converts it into a steam and this being lighter than gas stop when they leave the comet they stay in that position and moves in correspondence with the Solar system movement.

1. The Earth's Moon has no atmosphere or intrinsic magnetic field, and consequently its surface is bombarded with the full Solar wind. The Project opolow missions deployed passive aluminium collectors in an attempt to sample the Solar wind, and lunar Soil returned for study confirmed that the lunar regolith is enriched in atomic nuclei deposited from the Solar wind. These elements may prove useful reSources for lunar colonies.

WRONG

Here I go again. They say that the Solar winds travel at a tremendous speed; if this is So and our moon has no atmosphere how is it that even though the moon's entire surface has a large amount of dust. This we know for a fact, if this is So why do we not see any signs of these Solar winds that can travel up to seven hundred miles per hour. The information we have received concerning the moon's surface has shown puffs of dust rising from the inside of its craters near the top of the ridge. If there were Solar winds and the moon has no ozone layer or atmosphere then the dust would be like a sand storm, in out deserts multiplied by seven, this is only a guess. This fetches a problem. Where has the Solar winds gone? They say that our ozone layer protects us. And there must be a Solar wind as we can see the effect it has in our northern hemisphere. I do not think this is So. As there is no Solar wind affecting the moon? So I believe there is no Solar wind around the Earth's Hemisphere. The reason we have the northern lights is the gases in our stratosphere. If there were Solar winds around our Earth, then our Solar winds would not be dancing around like a butterfly on a ballerinas pirouette. With them being gases they would be blown to Timbuctoo. And the wife said I had trouble with wind. We now know that there are no Solar winds affecting the moons surface or the Earth's. The astronauts say that they could hear these winds from inside their space ship. Yet their ship was not affected. Am I missing Something? There is only one explanation they are not winds as we define winds they are Something else. What I have learned in the last six months of studying the universe I will try to use it. The only thing that comes to mind is they are like a wave lengths. The sun's rays are ultra violet waves we have micro waves and radio waves. So I will call them Solar waves. These waves like the others I have just mentioned need a form of a receiver. As a micro wave or a radio wave need them, then So must Solar winds. If these Solar winds are fully charged particles that are basically magnetically charged and the moon has no protection from them then surly the surface of the moon must have these magnetic properties. If this is So why are the space ships electrical parts not affected by them? The sun's rays are just as powerful as Solar winds, yet the Earth is affected by the sun's rays So why do the Solar winds not affect anything on the earth's surface. They say that our ozone layer protects our planet. Does it reflect them from reaching the Earth's surface why is there not a build up of these magnetic properties? There is another way of looking at this and that is extending Isaac Newton's law on gravity. If I have not enlightened, you up to the present time then I will let you into this fact later. If our ozone layer cannot protect us against radiation, gamma waves, and radio waves from reaching the Earth's surface how do they stop the Solar winds?

REPLY

If the Solar winds were reaching, and effecting our northern lights why are the not effecting our communications.

From Wikipedia, the free encyclopaedia
This article is about the stellar wind from the Sun.

Ulysses observations of Solar wind speed as a function of helio latitude during Solar minimum. Slow wind (~400 km/s) is confined to the equatorial regions, while fast wind (~750 km/s) is seen over the poles. Red/blue colours show inward/outward polarities of the heliospheric magnetic field.

The Solar wind is a stream of charged particles released from the upper atmosphere of the Sun. consists, protons and alpha particles with thermal energies between 1.5 and 10 keV. Embedded within the Solar-wind plasma is the interplanetary magnetic field. https://en.wikipedia.org/wiki/Solar_wind - cite_note-2 The Solar wind varies in density, temperature and speed over time and over Solar latitude and longitude. Its particles can escape the Sun's gravity because of their high energy resulting from the high temperature of the corona, which in turn is a result of the coronal magnetic field.

At a distance of more than a few Solar radii from the sun, the Solar wind is superSonic and reaches speeds of 250 to 750 kilometres per second. https://en.wikipedia.org/wiki/Solar_wind - cite_note-3 The flow of the Solar wind is no longer superSonic at the termination shock. The Voyager 2 spacecraft crossed the shock more than five times between 30 August and 10 December 2007.https://en.wikipedia.org/wiki/Solar_wind - cite_note-4 Voyager 2 crossed the shock about a billion kilometres closer to the Sun than the 13.5 billion kilometres distance where Voyager 1 came upon the termination shock. The spacecraft moved outward through the termination shock into the Helios heath and onward toward the interstellar medium.

The above say that the Solar wind is a stream of charged particles released from the upper atmosphere of the sun. This plasma consists of electrons, protons and alpha particles with thermal energies. If this is the case how can they change their speed. The sun's light is a of a Constance speed. Why then should these be So different in the Solar winds Within the five principal layers that are largely determined by temperature, several secondary layers may be distinguished by other properties:

The ozone layer is contained within the stratosphere. In this layer ozone concentrations are about 2 to 8 parts per million, which is much higher than in the lower atmosphere but still very small compared to the main components of the atmosphere. It is mainly located in the lower portion of the stratosphere from about 15–35 km (9.3–21.7 mi; 49,000–115,000 ft), though the thickness varies seaSonally and geographically. About 90% of the ozone in Earth's atmosphere is contained in the stratosphere.

The ionosphere is a region of the atmosphere that is ionized by Solar radiation. It is responsible for auroras. During daytime hours, it stretches from 50 to 1,000 km (31 to 621 mi; 160,000 to 3,280,000 ft) and includes the meSosphere, thermosphere, and parts of the exosphere. However, ionization in the meSosphere largely ceases during the night, So auroras are normally seen only in the thermosphere and lower exosphere. The ionosphere forms the inner edge of the magnetosphere. It has practical importance because it influences, for example, radio propagation on Earth.

The homosphere and heterosphere are defined by whether the atmospheric gases are well mixed. The surface-based homosphere includes the troposphere, stratosphere, meSosphere, and the lowest part of the thermosphere, where the chemical composition of the atmosphere does not depend on molecular weight because the gases are mixed by turbulence. https://en.wikipedia.org/wiki/Atmosphere_of_Earth - cite_note-19 This relatively homogeneous layer ends at the *turbopause* found at about 100 km (62 mi; 330,000 ft), the very edge of space itself as accepted by the FAI, which places it about 20 km (12 mi; 66,000 ft) above the meSopause.

Above this altitude lies the heterosphere, which includes the exosphere and most of the thermosphere. Here, the chemical composition varies with altitude. This is because the distance that particles can move without colliding with one another is large compared with the size of motions that cause mixing. This allows the gases to stratify by molecular weight, with the heavier ones, such as oxygen and nitrogen, present only near the bottom of the heterosphere. The upper part of the heterosphere is composed almost completely of hydrogen, the lightest element.

The planetary boundary layer is the part of the troposphere that is closest to Earth›s surface and is directly affected by it, mainly through turbulent diffusion. During the day the planetary boundary layer usually is well-mixed, whereas at night it becomes stably stratified with weak or intermittent mixing. The depth of the planetary boundary layer ranges from as little as about 100 metres (330 ft) on clear, calm nights to 3,000 m (9,800 ft) or more during the afternoon in dry regions. I was going to say that the above has educated me in the physics of the Solar winds. This would have me showing of a bit. It is So long winded to me that all I understood was the Solar winds are waves of charged particles.

When I mention gravity, I am not referring to the context they use today that was put forward by Isaac Newton. There is nothing in our Solar system or for that matter in space that draws thing in, this Also applies to the wording of stars imploding. These ideas are put forward by physitrists, scientists and all the other bodies that have an interest in space exploration. Let us look at what a theory is. This covers many explanations here are just few. I used the dictionary for these meanings to save my readers the trouble; Sorry l went on a wild goose-chase. None of them seemed to fit the description So I will use my definition. A crackpot idea, and all the other definitions given. I mean why use a sensible suggestion when you can use an extraordinary one that has no meaning. PhiloSophies this is a view point, idea, beliefs, attitudes this covers many things. I bet many of my readers could find a better theory than those put forward by the said experts. Even a class of school children, how about, a black hole = Goofy. String theory = Mickey Mouse and So on. This Sounds a lot more interesting like most professions they like to use words and phrases that we common people do not understand. This is that if we did know what they were talking about we would probably have better ideas on the subject and prove them to be incompetents. In years gone by us looked up to doctors as Gods. But know they have come to realise that the patient can in most cases now what medication they

RAINBOWS

    They say that the reason you can see a rainbow is that the sun's rays are hitting the raindrops that each colour is produced from the sun's different colours. So on a rainy day where there are trillions of raindrops in a given area. And none of them can be seen how this is possible. If you take what the sciences and physics say that this is because the sunlight hits the raindrop at a given angle. Why can the rainbow be seen by different people at a different angle when they are standing at a different part of the Earth. Surly if this was So then Some people standing and viewing the rainbow it would not show as clearly as a rainbow. There would be a deviation in each of the colours and they would gradually lose the basic colours. The scientists say the rainbow can be seen because the sunlight hits the raindrops at a forty two ^{0} angle.

    Has anyone thought about this, what is the chance of all these trillions of raindrops being in the correct position. There are seven colours, the mind boggles at what this mathematical total could be. What I am going to say next is going to increase this mathematical sum into the deudecillions. Come on mathematicians have you thought what it could be that would increase this into deudecillions, and probably more. I am no maths genius but maybe they could work it out. Here is my question. They say the rainbow can be seen because the light hits the raindrops at a certain angle. It is such an easy way to work out this problem. So these duedecillions of raindrops are at a certain angle. But every split second the next lot of droplets take the place of the other descending raindrops and So on. Let us stop the frame…. Okay that's long enough. Now what do we see. The sun's rays are hitting the raindrops at forty-two degrees. The raindrops above and below these raindrops are such a minute distance away from each other there should be one or two things that should happen. Either the different colours should be wider with no effect on their colours, or they should be more of a paler shade. If you look at the colours in a rainbow you will see between each different colour that there is a slight paling of each of the colours. We can see this is a fact. Can we really be expected to believe that the sun's rays which are shining through space that there are gaps between each micro wave that it would leave blank spaces. This tells me that when it is raining and the sun has risen to the position that there was only a fraction of the sun not above the horizon. The sky should be one continuous rainbow. So there must be another Solution. Where else could the colours come from? The physicists and the scientists say that the colours come from the white rays of the sun, Sorry I disagree. The sun shines with a white glow. The white rays do not make the colours of the rainbow. The seven colours of the rainbow (gases) are in our atmosphere. The different colours come from different gases that are in different places So his means that they are not in uniform. Also when they are at the first stages and Also when they are rising they will be intermingled by the wind. The raindrops must contain the seven different colours would be imposable for each one to contain only one colour. There are So many reasons why rainbows are seen as in a bow I cannot see with all the reasons that physics give does not come into line with my logics. Or better still my physics. I do not see why there is any reason why I cannot call myself a physiatrists as the number of students that hold a degree in physics is only one third. This is according to the Sun newspaper. Saturday, July the sixth. Page six. Let us go get back to the rainbow's configuration. We must take it that each raindrop contains the seven colours. The raindrops on the furthest left and the furthest on the right would be at 42 degrees So why are the raindrops at the top centre still showing up as red as these cannot be at the same angle of 42 degrees. The figure given is forty two degrees if this angle given it must be a precise angle. I am talking a precise angle. So this angle must change in respect of the far ends of the rainbow. I am no mathematician that can deal with angles So precise, they started it So let them tell me how they have worked this out. Of all the wonders created by our LORD to me this is the only one that outstrips all of the others. Not in its colossal affect but in its ability to have this affect degrees So accurate that how the experts have worked it out boggles my mind. There is a scientific explanation for all of the miracles in the bible, but none can explain why a rainbow is the way it is. If the angle is 42 degrees, then the probability of these being in a bow and not following the contour of the Earth is beyond me.

The boffins say that the white we can see is a combination of the colours. If this is the case as all the colours are in our atmosphere why can we see only the white? When I was at school and during artist class I never produced a white colour? I have mixed many colours when I was starting to decorate and never got the colour white. If you look at the sun you will see the rainbows colours around the circumference. This is not recommended find pictures of the sun in one of our universe on the internet. The reason we cannot see them is because sunlight is stronger. Why is it we can see these colours in a rainbow? This is because the sunlight strength is reduced by the raindrops. How is this for a thought, the rainbow can be seen at 42 degrees to the observer. If this is the case then am I the only one that can see a flaw in the physics of this.

HOW PRISMS WORK

A triangular prism dispersing light waves. Light changes speed as it moves from one medium to another (for example, from air into the glass of the prism). This speed changes the light to be refracted and to enter the new medium at a different angle (Huygens principle). The degree of bending of the light's path depends on the angle that the incident beam of light makes with the surface, and on the ratio between the refractive indices of the two media (Snell's law). The refractive index of many materials (such as glass) varies with the wavelength or colour of the light used, a phenomenon known as dispersion. This causes light of different colours to be refracted differently and to leave the prism at different angles, creating an effect similar to a rainbow. This can be used to separate a beam of white light into its constituent spectrum of colours. Prisms will generally disperse light over a much larger frequency band width than diffraction gratings, making them useful for broad-spectrum spectroscopy. Furthermore, prisms do not suffer from complications arising from overlapping spectral orders, which all gratings have. Prisms are Sometimes used for the internal reflection at the surfaces rather than for dispersion. If light inside the prism hits one of the surfaces at a sufficiently steep angle. Another thing you must fetch into the equation is that to me it's must be Something else, but the gases that are causing this phenomenon. There are too many contradictions to know which one is right. One that does stand out is why Earth's gases show in a bow. This does not correspond with the way gases are supposed to form. If it is the angle that makes it possible to see the rainbow at what angle does it make it impossible for Someone to see the rainbow? In all the miracles that happened the scientist and physicists say that these can be due to natural forces.

You do not need rain to have a rainbow. So that means that when we have a rainbow when it is raining it is not because the raindrops are at a certain angle to the sun's rays this is what I have always suspected as the combination of this accruing is beyond belief. All sun's show rainbows around their circumferences and there is no rain. That means it is gases that produce the rainbows. So do the rain drops hold the gases in their air spaces? Remember that the heavier the gases then they would be the ones at the lowest level. Now comes the question why in a bow? They should be in a horizontal line. Or on the same contour as the Earth not between the two. Remember these gases are all rising all over the Earth If you heat metals up here on Earth you will see the colours that these gases produce. I am dying to see how the space establishment react to my logics. This is my observation that the rainbow should not show as a rainbow. If you think about it every rainbow is seen at a different angle. Let us say we are looking at it at a 90° angle and we see a rainbow this means it is reflecting the light at a 42° angle, this is the angle that the scientists have said that it reflects the colours. This to me means there is no deviation on this angle. A perSon next to you will see the rainbow but the colours will be of a different hue. There are many different views on rainbows on the internet I could download these but there are So many and they all contradict each other that I do not want any of my readers that do not want a universal degree to must go through this ordeal. Those that do need this information as I have said there are many internet sites that you can go on. I hope the information I have taken from Wikipedia will give you enough information. A rainbow is not located at a specific distance from the observer, but comes from an optical illusion caused by any water droplets viewed from a certain angle relative to a light Source. Thus, a rainbow is not an object and cannot be physically approached. Indeed, it is impossible for an observer to see a rainbow from water droplets at any angle other than the customary one of 42 degrees from the direction opposite the light Source. Even if an observer sees another observer who seems "under" or "at the end of" a rainbow, the second observer will see a different rainbow—farther off—at the same angle as seen by the first observer. Let me put it another way. Sometimes I see a rainbow, let me say for arguments sake it is at eye level. The next time I see a rainbow it is above eye level then surly I am not seeing it at the same angle So how can

I see it. to look at this in a different light, (I like that pun,) if I am looking at a rainbow directly and I see all the colours at that point that means that I am seeing the colours reflection at forty two degrees. I am Also looking at the colours at each end of the rainbow then I am not looking them at forty-two degrees

Rainbows span a continuous spectrum of colours. Any distinct bands perceived are an artefact of human colour vision, and no banding of any type is seen in a black-and-white photo of a rainbow, only a smooth gradation of intensity to a maximum, then fading towards the other side. For colours seen by the human eye, the most commonly cited and remembered sequence is Newton's sevenfold red, orange, yellow, green, blue, indigo and violet. The number of colours in spectrum or a rainbow is red, orange, yellow, green, blue, indigo and violet. Isaac Asimov does did not include indigo as a colour as all this was, was a dark blue. With all the information on the internet there are not two of these So-called experts that have come to the same decision. What does this tell you? Simple they have no idea. That to me means only one thing there must have been a creator. Don't go huffing and puffing you will have my explanation later? need to see if I have explained how the sun's rays picks out all the colours on the same plain as each raindrop contains all the colours

A prism reflects the colours. A prism shows that the white light has Somehow vanished and we only have the colours remaining.

How then do we still receive a white light from the sun?

There is another theory going around and that is they are picking up light from gases that are exploding in the Milky Way. I am now back on earth. I am not going to mess around with any thought of using explosives or petrol to do this experiment as I would probably blow myself up. What's that I hear, do it, do it? I will use a bit of knowledge that I have about explosions. You have an explosion there is a flash of light you are blinded by this flash, so you automatically close your eyes. Let us use another scenario. You are expecting an explosion, So you close your eyes in anticipation of a blinding flash of light

Let us wait a bit. A bit longer. You open your eyes. There is nothing, So now we now that here on Earth after an explosion the light has ceased to exist. Now why should anyone with an ounce of knowledge think that because it has happened in space that it should react differently? (I am not talking about sunlight as that has a continua's power sauce. If it reacts the same as it happens here on Earth, then we would see no light. Now we come to what the physicists are saying that the gases will explode and this will created a sun. When gases explode what they create is another form of gas. You can see the gases produced by the sun, So why is it that when the sun has producing these gases and the tremendous heat that the sun produces these gases do not explode. We know that logically they must be too far away from the sparks. So how can gases explode in the universe where there is nothing to ignite them.

I am glad that the authors are putting forward another theory that instead of a projectile being the size that would take the core of the Earth into making the moon. Surly the way things are going the moon should be bigger than the Earth. But being sensible chappies, they must have realised as I did what the outcome would be. So, they come up with suggestion that it was only a bit more than the size of the Earth. As I have said this would have destroyed the Earth.

Another scenario of mine can Also be used in the context that the object hitting the Earth hit it in a different direction. This occurred at the beginning of Earth's formation. Currently there was no ozone layer to protect it. So the object would arrive at the Earth's surface in a more direct direction. This in fact would create a more circular and would definitely destroy any matter it hit and Also destroy itself. Any that did enter space would have the same scenario as the previous statement. I consider that blows all their theories into space. Let us presume that this Mars object hit the Earth at an angle. It would must be at a precise angle. It would Also must the correct weight, and the speed would Also come into the equation. If one of these theories was out of proportion it would affect the size of the moon being created. Now if we Also put the gravitational pull of the moon on Earth's tides then it would must be at the right distance. There would Also be a very wide channel gorged out of the Earth's surface. This Mars chunk of granite was half the size of Mars.

Do the scientists and physicists now how the universe works well I will remind them of what they said. All matter in the universe is travelling outwards. Maybe in their later books they can explain to my readers and I how this moon debris has a mind of its own and is contradicting all that we are told by scientists. To the above scenarios and tests done to find out where our satellite the moon came from. The first scenario that is put forward is that an object hit the Earth. The planetary object struck when the Earth was just about fifty million years old. Formed – a "proto-Earth," as scientists call it. This then makes the Earth half a million years older than the moon. They should both be the same age. The object that is supposed to have smashed into the Earth would not have carried on into space; it would have embedded itself into the Earth's crust.

God created both at the same time So let's do a bit of calculations and see which one is right. From the information above this collision would appear to be a direct and not at an angle. As I pointed out there is no sign on earth of a wide long trough. To replace the amount of matter needed to form our moon which they say is one quarter of the Earth's size. It must be a big object. An object scoring the Earth would lose a lot of its matter. Let's look at it from the other angle. What would happen to the Earth's matter it would not come out in a large enough size they Also would be smashed into smaller chunks? If we can recall what was said earlier, it was that all the debris that came from the Earth's surface came together and this caused debris to be thrown into space So creating the moon. What I know about debris entering space, it does not congeal together. Once it enters space it then comes under the influence of the sun in that Solar system/galaxy and as with all other objects they travel at fifteen thousand miles per second in an outward conjecture. So why should our moon act in a different way to all the other objects in our universe. Another point I would like to put forward is the moon is a quarter the size of the Earth. the object that was supposed to have struck the Earth 4.5 billion years ago and removed matter from as deep as two miles, this to Einstein's mathematical equation means that it was five miles from the centre of the Earth's core. We must stay with this time scale, at the very beginning if we look at the lower regions of the Earth then this consisted of molten plasma. So in the short time from the forming of the Earth the earth must have had a very deep crust that went down to five miles from the centre as this is what the physicists and scientists are saying? If this was the case, then Earth should be one Solid block as the moon is. I mean it looks like the Earth's surface has not changed in the last fifteen billion years. This should give a lot more theories for the experts to come up with. I think that my observation on what the physicists and scientists say that they will must think again about how the moon was formed. If you have a force that is sending matter in an outward conjecture then there is no force that can bring this together. And if you know my observations about gravity this does not draw things in to its centre.

A BIT ABOUT ME

I am no physicists, mathematician, or had any high education. My education consisted of my attendance at a primarily secondary school. That was when I went. In the early days at junior school owing to illness my education was usually interrupted. Not being able to revise or never getting home work when I finally went back to school I didn't have a clue what the teachers were talking about. This of course made the bottom of the class in all subjects. I am glad that there was no basement at this school otherwise I would have been the only one in that class. I came to the decision that all my mates would have said I was the bottom of that class but being an optimist, I looked on it that I was the top of the class. My parents decided to move to a new house I was hoping this was not because of my poor educational skills and they were ashamed of me. That can't be true because when I came home my mum had told the neighbours where they had gone. This made me realise that they wanted me to follow. If they really wanted to leave me and go Somewhere that I would not find them they would have gone to Milton Keys and not to China. To get back to my school days I don't know how it happened I think it was because I was older and not So prone to illnesses, but I had to attend another junior school. I was puzzled because all my previous school mates where in a lower class than me. You might have thought this was a good thing. Your wrong what it meant was I was back to square one. These new class mates were far more advanced than me, So here we go again I was back to the bottom of all the classes. So from twelve and a half years of age to thirteen and a half I had had enough. So I stopped going. I might have decided on this decision but don't get me wrong I couldn't hang about the streets as I would have been the only child playing by myself, and I would have stood out like a not going to school. I got a job working on a farm. It was twelve months before they caught up with me. I left school at fourteen and a half and got a job on a farm. This was ideal as you didn't must do any maths as the saying goes don't count your chickens before they hatch. Before this came about I thought I had better look for another job. You won't believe what job I tried to get it was as a bus conductor. How's that for being an optimist. As was to be expected I failed the maths test. it was a good thing that the maths test was conducted by an instructor that said I had only just failed, and he gave me a maths book and he said apply again in eight weeks' time. The book was brilliant it was So easy to understand. I returned to pass the test. The way the pay was worked out was very complicated. There were So many variations, basic wage, time and a half, double time, spread-over's, and were in different combinations. The other employers would come to me to work out their wages. After being So thick at school I felt good that I could help my fellow workers. Thanks to the internet I have been able to understand Some of it. My Son came down to see me and we were discussing the effects of gravity. My logic was that gravity does not draw things into its orbit. It is the weight that makes it fall to the surface. He put forward the difference in weight when Someone was on the moon to their weight on Earth. Owing to the spinning of each planet this made a difference to the weight of that perSon. I thought of this and came to the decision that I was going about this the wrong way. I had come to the decision that planets objects were not controlled by the gravitational pull on the object but by the weight of the object. If we look at the Solar system, note that the planets and asteroids do not fall towards the sun. This means that my logic is they are doing the opposite. They are keeping that object in orbit and the weight of that object does not make it fall. This now puts a different light on matter weighing billions of tons on a table spoonful. My new word is Penny Force. This will be used in place of gravity. This is to make it easier for all those who are experts concerning the universe. We all must learn So don't blame me I have had to start ratio of planets anti gravitational pull. The images below are explained straight after the images. It will become apparent after reading about them that the three images are the possibility of the shape of the universe. Let me explain to everybody which one it is and how I have come to this decision. The theories put forward that dark matter and dark energy is Some way controlling the universe. Poppy cock. I will put them all straight in this matter as could anyone that has a small amount of knowledge about space, If it was not for the influence of the stars gravity then all the matter after the Big Bang would be heading for the blue yonder. But because of the sun's gravitational effect, not in the sense that gravity pulls thing into its domain, but it stops matter from leaving the universe. This does not in fact stop matter from going further out into space owning to the expanding universe I realise that none of these images are correct in their theories. You might as well ask why? 'Why' I am glad you asked. Erm, erm. erm Got it if you know me you will realise that I very really use theories. So here is my logic on why the above theories are wrong. As most of you know that have a little knowledge of the Big Bang and how the universe works will not need me to explain, but trying to keep it in plain English for any new readers that wish to learn about this subject and not be at a loss as to what the hell are they talking about. I will attempt a short version. The universe is spinning this is in a sphere, and all the Solar system are in this sphere. Our Solar system is in this sphere. So why should anything in this sphere change its shape. I am not worried whether it is black holes, dark matter, gravities, or any other fancy theories that they can come up with. I hope that can be understood by all my readers. The information below has my head going around and this must apply to my readers who have no knowledge of the universe and all the jargon. Most of it I will disregard and use plain English as much as possible. General relativity describes how space time is curved and bent by mass and energy. You have more chance of bending an iron bar. Nothing curves space otherwise how can anything be bent when there is nothing there to bend. If space was bent then the planets, sun's, and asteroids would not be heading towards the outer regains of the universe. This would mean that this mass and energy was in fact stronger that the universes ability to spin the universe around. The topology or geometry of the Universe includes both depending on whether Ω is equal to, less than, or greater than 1. These are called, respectively, the flat, open have heard of hedging your bets but I think if they could have used just one of them.

    Ninety nine years after Einstein's theory that time and space bend, a $700m NASA probe has proven that half of this is right. The part that they have experimented on is the geodetic theory. This is the bending of space. They are now working on the other half. That time can Also be bent. I dread to think how much this is going to cost, if they read my observations on the bending of space and time it might save them Sorry us Some money. I reckon on giving a low estimate of $700m it will be a few dollars short. The physicists have agreed that Einstein was a genius or maybe only half a genius. I consider that even though he only got it half right I bet he did it for a lot less than $700m. Maybe it would save us a lot of money if we just took Einstein's work as a fact. I mean to say is it going to be the end of civilisation as we know it. We are supposed to be intelligent. Surly all this money would be better spent on the starving and the needy, NASA's results from Gravity Probe B, this was in the late nineteen fifties confirmed that the results prove that the possibility of there being a mistake was that remote it was not worth mentioning, and that Earth had the ability to distort the fabrics of space. Going on what all the experts say is that the fabrics of space are like a hidden fabric in space. Why they keep coming up with these invisible fields I think it is just to confuse us mere mortals. Let me keep to the story So far. Earth is distorting space. If space was being distorted that must mean that there were parts of space that were concentrated and at the other extreme it had more space. I am lost how can nothing be messed with. Maybe like the ordinary perSon I am getting it wrong by the concept of bending. Where is it bending from and where is it going. I am going to get right to my point about space bending.

RUBBISH

If all you budding physicists, and Star Wars fanatics thought the bending of space was confusing then you are about to have your mind So confused that you wish that you had got more of an interest in ghosts, spirits, aliens and mystic Meg who could read your stars.

If what they say about space bending and the effect of gravitational fields as with all these theories and you will note I am putting the NASA probe into these theories, then as with all of these I will do as I always do I fetch things down to Earth. We have space here on the Earth. That means that all the things flying through space is being bent. I will must get in touch with NASA and ask them if they could passably send me a bit about their experiment on how to see space bending. When I get this information, I will go outside and throw a ball and see space bending. What did they see that was happening to the space that was next to the bent space? Then again if it is like space the bend goes on forever. It was decades after the NASA probe before any more investigation was done on trying to Sort out this puzzle that's the lot about the bent space. We will now go on to Sort out the bending of time I know about bending time once I dropped my watch and it has never worked since. Let us be serious about this it's not Something you should joke about. There is going to be a lot of money spent trying to figure out this problem. The below information comes from various universities and persons probing the information that NASA and Einstein's work on relativity. This information that they are using is passed on work done by Einstein. Francis Everitt, the Stanford University profesSor who has devoted his life to investigating Einstein's theory of relativity, told scientists at the American Physical Society it would be another eight months before he could measure the 'frame-dragging' effect precisely. Understanding the details is a bit like an archaeological dig,' said William Bencze, programme manager for the mission. 'A scientist starts with a bulldozer, follows with a shovel, then finally uses dental picks and toothbrushes to clear the dust away. We're passing out the toothbrushes now.' Well it looks like we are going to get a clean view The Gravity Probe B project was conceived in the late 1950s but suffered decades of delays while other scientists ran tests corroborating Einstein's theory. It was Everitt's determination that stopped it being cancelled. The joint mission between NASA and Stanford University uses four of the most perfect spheres - ultra precise gyroscopes - to detect minute distortions in the fabric of the universe. Everett's aim was to prove to the highest precision yet if Einstein was correct in the way he described gravity. According to Einstein, in the same way that a large ball placed on an elasticated cloth stretches the fabric and causes it to sag, So planets and stars warp space-time. A marble moving along the sagging cloth will be drawn towards the ball, as the Earth is to the Sun, but not fall into it as long as it keeps moving at speed. Gravity, argued Einstein, was not an attractive force between bodies as had been previously thought.

Few scientists need the final results, which will be revealed in the future, to convince them of Einstein's genius. 'From the most eSoteric aspects of time dilation through to the beautiful and simple equation, e=mc2, the vast bulk of Einstein's ideas about the universe are standing up to the test of time,' said Robert Massey, from the Royal Astronomical Society. He said the mission was 'legitimate science' to test a theory and confirm its brilliance, but others have criticised the costs and length of the study, claiming that what was announced had already been shown. Sir Martin Rees, the Astronomer Royal, said the announcement would 'fork no lightning'. This will come as a surprise and then again it shouldn't be as you know my view about it. And it's the cost

THE THEORY EXPLAINED

When Einstein wrote his space and general theory of relativity in 1915, he found a new way to describe gravity. It was not a force, as Sir Isaac Newton had supposed, but a consequence of the distortion time, conceived together in his theory as 'space-time'. Any object distorts the fabric of space-time and the bigger it is, the greater the effect. Sorry Einstein.

RUBBISH

Why do these physiatrists keep finding other physiatrists wrong? Einstein says that it is the bending of space and time. Now I will jump on the band wagon and say that Einstein has got it wrong. As far as space-time goes there is no such thing. The above puts forward Einstein's view that objects bend. How can anything that is travelling through space at fifteen thousand miles per second have any effect on the matter around it. Space is a vacuum any object going through a vacuum cannot affect anything when there is nothing there. I am Sorry, but I am going to put forward a theory. And a experiment that applies to it being done here on Earth. If it works, this will show that there is more of reason that the same scenario will happen in space. Let us drop fifty p from a plane. Stop and think what you are doing. The expense of this experiment is too large just because it is tax payer's money we do not must go mad. Let us drop one pence from a hot air balloon. If anyone can find a way to drop it from a kite this would be cheaper. With Isaac Newton's theory that equal forces balance each other out this should have no effect on the surrounding space. How's that for a theory. What has the bending of space got to do with gravity? So my decision is. "No chance." Just as a bowling ball placed on a trampoline stretches the fabric and causes it to sag, So planets and stars warp space-time - a phenomenon known as the 'geodetic effect'. A marble moving along the trampoline will be drawn inexorably towards the ball. This is the same as the above Einstein's findings, or am I missing Something. The first theory says that an object distorts and the bigger the object the more it bends. The next gives a marble on a trampoline; to me this is the same theory. They Also make the sun as part of this warping. The sun does not distort space in the case of warping it. The sun is the centre of each Solar system. This means in fact it cannot bend space as it is not moving all other objects are moving away from it. How is it possible to compare space in the same breath as a trampoline? Thus the planets orbiting the Sun are not being pulled by the Sun; they are following the curved space-time deformation caused by the Sun. The reason the planets never fall into the Sun is because of the speed at which they are travelling. If space was bending, then this would have shown by photographs from the past. All the planets are in the places they should be. If these objects were being affected by warping then as there are different sizes and masses, then they would have been affected differently So they would be in a different place in the universe. At the point where the stretching was at its weakest then the heaviest would follow the contour. Let me use their cloth theory. They put forward the idea that we have a sheet of cloth. We roll an object down this cloth and it changes the way the cloth looks. At the centre of this cloth it retracts it if this is how it works then in time the lighter objects would Also go to the centre of the cloth. If we think about this then we conclude that all material will finish up as one big clump of matter. This of cause would include planets, asteroids and then the outcome of this would be that they would burn up under the heat of the sun. According to the theory, matter and energy distort space time; curving it around them 'Frame dragging' theoretically occurs when the rotation of a large body 'twists' nearby space and time. It is this second part of Einstein's theory that the NASA mission has yet to corroborate. Well what I can say about time bending. I have no way of contradicting their theories, if they have not got any yet. So I will just keep to my own logic "rubbish." You cannot bend space in the sense of space itself. The only way we can relate to space is from the beginning of the universe. This is the only way we can see time. Before this occurrence there was no time. When you look up into space that is how old space is as you cannot do anything with nothing it does not matter what you try to do. The only one that has ever been able to do this was *our* LORD. If it was possible to produce Something from nothing this Earth would certainly be a different place. Global warming would stop. The mind just boggles. I hope if the physiatrists can come up with a theory.

Gravity as used by physicists to explain the reason why every planet and their moons have the ability to draw thing in. this is not the case it is the movement of the galaxy that holds things together. The Milky Way has many galaxies and Solar Systems this does not put it in its right prospectus. The Milky Way has billons of galaxies. All the Solar systems are as you would expect are in a uniformed position around its sun. And they are all in this sphere known as the Milky Way. Earth is outside this band why? The position of the Earth had defied the law of nature by this I mean the way the universe works. The universe was expanding in this uniformed way in all directions. Now we have the universe expanding, there are no changes over billions of years. The physiatrists and the scientists say that the galaxies have all descended from the Big Bang. They say that there are two differences, one is the Andromeda galaxy and the other is Earth. These were created at a different time to the Milky Way. That to me would mean that they are putting this as the same time as the Earth arrived on our planet. We know by the expert's deduction that Earth was the second of Solar systems. As you look at it, it appears to be outside the Milky Way galaxy. Now why would there only be two scenarios one is a galaxy and the other is a Solar system. And there is another thing that would stand out like a moving car on the M25. That is the Andromeda galaxy. This galaxy is part of the Milky Way and was created by the Big Bang. Now Earth is a difference prospect altogether. This is the only Solar system created from the second gene ration of Solar systems. Now I ask myself, How is it possible for there being only one Solar system being produced The universe has had billions of years of existence and surly there would have been more than one sun going into melt down. This is only a saying as they would have exploded. You might ask how is it that a galaxy that is the biggest in the Milky Way came about. Was this the first bang?

Solar energy and where it is going. As the utility firms invest more into remunerable energy the price of gas and electricity will become cheaper. The price has already decreased in these commodities. (2016.) what effects will this have on the house holder that has Solar panels installed? There is only one outcome and that is as the price of these insulations become costlier it is of no benefit to have them installed. The only way that house holders would find a benefit if the Solar panels could be installed in an easier way, thus cutting down the price. The time must come when the lowest price that the utility firms can sell their product reach that price. By this time I would think that Solar panels had reached their maximum efficiency. The modern Solar panels now produce more power than the first panels could do. As it will show in time there will be more innovations in the Source of power and were we obtain it from.

There was nothing. If this was the case well then when perform miracles the first thing you would do is create Some and how did our God appear? This question is the most outstanding question that is far the hardest for us mortals to understand. If space consisted of nothing, what possibly could create a God? We know that the past like the future goes on/back forever. We know in our universe and especially on Earth that there have been miracles. So the question is at what time it was in when our LORD appeared. Should we must look at the time just before the universes arrival? What if it was and there was nothing. Has our lord been here as far back as you can go back in time which is for ever? I can not see this because surly he would have created the first sun before this. if this is not So then how did he appear. Whatever it was surly if you could thing. Space as we now it is at absolutely zero. I cannot see what can be created in a temperature of this level and then why not as our Lord is everywhere.

If anyone can put the finger on this conundrum then this would be revolutionary. That would be the miracle of all miracles.

How about this question that no one has thought of answering and that is. If there was no universe, then what would be the use of pace?

The physicists and scientists say that sunlight bends. I do not know what scientific experiments were carried out to come to this decision, or how the physicists, astrologists have come to this decision or anyone else looking up to the skies. There is a simple experiment that can be done and it puts this theory and the theory of black holes into the lime light. I have no equipment, telescopes, or any other form of equipment to do or see what is happening in the universe, all I do is read all about the theories on the internet and put logic to them. Let me explain how my experiment works that proves that the sun's rays do not bend. Place yourself in a dark room. It is recommended that you now your way about this room. You turn on the light. Then you open a cupboard door, you will see that the inside of the cupboard on all three sides are lit up. Now you stand back in amazement. I am amazed you say, or words to that effect. Nearly all the inside of the cupboard has lit up. This in fact means that the light has not only bends but it has bent one hundred and eighty degrees.

Let us start at the beginning when I say the beginning I don't mean the beginning and will tell you why later (at the end of book) I mean the Big Bang. The astrologists, scientists all agree that this is a fact. That means the universe is a fact. Now we come to a lot of propositions this is concerning life. All the planets, asteroids and gases from the first Big Bang have been repeated and each galaxy was formed. So over the last five billion years they have all gone through the same stages as they will keep doing. What are these stages you might as well ask? The original sun exploded. Let us all agree on that. There seems to be more than one opinion on how this accrued. (I will go into that later.) What we have is hot plasma, Solids and gases being blasted into space. This is the basis of all matter in the galaxies. As Soon as they were travelling through space they would all be affected in the same way? There is only one part of these three elements that did not change, and that is the gases, the physicists and astrometry have come up with the theory that gases in the universe have collided and formed new sun's. For this to have happened meant that there was a gravitational pull. Gases do not spin and therefore they do not create a gravity field. And if they did they would still not collide. I am Sorry but the answer why they would not collide will be explained later. Gases are usually found around the circumference of planets and sun's. The reason for this is that the gases are produced by the heating of metals. Why then are the gases all approximately in the position they are in. As is my wont later I will give the answer. What makes gases show their colour? We know that different metals give off different colours. How then are the remaining two objects affected by space? Most readers will, probably know the answer. It is the temperature. This plasma is travelling through space that has an absolutely zero temperature. This then cools the plasma until it forms a crust. And then bang. We will go back to the way Earth reacts to the universe. We of course could go the same way. There is a difference man has increased the temperature of the planet, and the bible says that the time will come when fire and brimstone will rain down on the earth and he will intervene this from happening. There are a few points that we must notice about this statement. One is why can a statement like this be made. We in our present day know that absolutely zero temperature cools everything down. But if we keep increasing the temperature of our planet what do you think the scenario will be. The opposite will happen to the effect of zero temperature. This goes against all the physics of the universe. How could it be possible for Someone to know this at the beginning when the bible was written?

EVOLUTION

I have heard all the talk about the missing link between the ape and man. Where are the missing links before apes and chimpanzees? Why they want to keep on about this factor I do not know. There is Something much more important where the missing link is more noticeable. Or should I say not notable. It's the human brain. You might have evolution were it can be seen in the gradual changes in animals, their body and habits change but the difference in the brain do not change the only way that animals evolve is through watching other animals whether this be the human animal as they have encroached on their environment. That is why mostly apes and chimpanzees have picked up are habits not the opposite. Apes are supposed to be our nearest relative if this is So then I will start to think this is right if it can do one simple task for me. Find the beginning of the cello tape. I cannot find it So am I still a primitive species. This missing link is not one missing link but trillions.

All the answers in this book?

THE BIBLE.

Evolution which to believe.

In my calculation don't quote me on this, but you would need to have a one billionth of a change for every year from the start of the universe. That is fourteen billion years. Since the discovery of DNA and how it is used especially in birds and insects Also have a DNA this Also applies to plants. Seeing that we were all created from the dust of the Earth then this is no surprise. Evolution must start from Somewhere but where did it start? With all the theories that keep getting bandied around there has not been one that has convinced me that, that is how it all started. It is only reasonable to think that even with the evolution theory that we originally came from bacteria as this must have been the first sign of life on the planet. It was the discovery of the structure of deoxyribonucleic acid, and the technology to sequence the genomes of both humans and animals that our DNA has a lot in common. The most usual story to connect us to animals started from comparing us with our nearest look alike animal the ape. To brain-wash us easily led animals they put in words like our nearest relative is the ape, and then they have come up with another brain washing scheme and that is our cousins. I think they have missed a few good ones how about, brothers, sisters, mum or dad. I wonder how people would react to these two expressions. The difference in the DNA between Humans and animals is So minute I don't know why they bother to say there is any difference. I have another one that we can look at and that is the between us and an ant the difference is around about the same. So why do we not call them our nearest relative, or our cousin. I won't name the other two but you know where I am going with this. The biologists have discovered dinosaur bones it is no surprise to find that we have a lot in common with our animal friends. How much humans have in common with animals may come as a bit of a shock. While it is understandable that we share Some DNA with our cousins the apes, we Also share Some DNA with other less similar animals on this planet. These differences when compared and not wishing to be contradicted I will say around about ninety percent to near ninety nine percent. Because of these similar traits then we have learned that they can be used to our advantage. There are many things that we have found out; we do a lot of test on mice. Humans and mice share nearly 90 percent of human DNA. This is important because mice have been used in laboratories as experimental animals for research into human disease processes for years. Mice are currently used in genetic research to test gene replacement, and gene therapy because they have similar gene types to those of humans and will have similar reactions to diseases and disease processes. Also there was a fallacy at one time that animal diseases could not be passed on to humans. These happenings are occurring because we have taken them into our environment. There's TB, bird flu and mad cow disease to name a few. There are many animals being tested to see if we can learn anything about similar diseases. Dogs share about eighty four percent of their DNA with humans. They are studying eye defects to see if it will benefit humans. I would imagine that is not the only thing they are studying. There are living things on our planet whether it be animal life or plants there are parts of their construction that have a valued use to mankind it is just it. Chickens are a different thing. Birds are not in the same field as humans. Their DNA cannot be anywhere near the eighty four percent of animals. Going by my logic if the chickens have a disease that can be passed on to humans then the vaccine that can cure a chicken of flue then there is no reason why that vaccine cannot be used on humans, maybe with a small adjustment. (Man must realise that most plants and animals have a part of their construction or the way that they survive in their environment can be used to our advantage.)

ICE AGE

*https://en.wikipedia.org/wiki/**Ice_age***

There have been at least five major ice ages in the Earth's past (the Huronian, Cryogenian, Andean-Saharan, Karoo Ice Age and the Quaternary glaciations). Outside these ages, the Earth seems to have been ice-free even in high latitudes. What does this tell me about the Earth's life in the universe. As it is stated there was five ice ages. Now let us look at another thing it tells us. Between these times there were ice free places at high latitude. This to me means global warming was at a really high temperature. Did this period mean that the Earth was in a serious drought condition? If So was it serious enough that everything perished. What we do know about Earth's weather is that when the coldest regions become colder than the hottest regions become hotter. I have just noticed this So instead of having an ice age and a drought age we must put the two together. This then reduces the periods of destruction of the Earth. Let us presume by the Geologists findings that it is six periods of the ice age. Now we take the floods. We know of one and that was during Noah's time. Now the bible says that our LORD said that he would not send any more floods because of man's indiscretions. That tells us that it must have accrued before and the bible will confirm this, it happened twice. Now let us look at volcanoes not to close otherwise we might burn our fingers. There has been many but going by the eleven strongest there has not been one that could have destroyed the Earth. We are always hearing how volcanoes are going to destroy the Earth. There is one that is below Yellow Stone Park that is said to be one of the most powerful if it was to erupt. This being So it wouldn't destroy the Earth. And there is a long way to go before the Earth's crust is So thick that the Earth's time was up and bang we all go. This is what happens when a planet has a molten core. And if you go by what the bible says and what happens to this type of space object there would be an increase in volcanic activity. When this comes about our LORD says he will intervene. Surly this is a logic thing to presume. Never mind let's just go with the five ice ages and what it tells us. We know by the geologist's findings that in five places they have found ash in the regions that Sodom and Gomorra were said to be. But as we know these were iSolated cases. So we cannot include them in our study what we do have taking the lowest number of occurrences, is six ice ages and two floods. You might ask. What is the point of this study? This study shows that the total times that the Earth was destroyed were eight times. Now this is a blinder. There must be a creator. "Why" did I hear Somebody ask why? Well just think about it every time the Earth was destroyed. You will must give me a minute as it is So obvious, I cannot see why nobody has not thought of it before. Are you all sitting on the edge of your seats well here goes? Every time there is a destruction of the Earth then we must start all over again. There are people on Earth that believe that we originated from outer space. Don't laugh at me laugh with me. This is suggested by one perSon but then it starts to accelerate. And in time you have a Society who believes in this. They then go through the process of brain washing these easily led people to send them money. What we have here is planet of aliens looking down on Earth and deciding they will send down Some microbes, plants and insects. After a while Earth has an ice age. There is an emergency meeting on planet Zog There has been an ice age on planet Earth we will must reinstate what has been destroyed by the ice age. 'Let us send down a batch of our planets reSources.' It was not long after this, that another ice age overcame the planet Earth, and all was destroyed.

To those that have the theory that we have come from another planet. Below is my interpretation of what would have accrued on this planet and the aliens discussing their planned landing?

On planet Zog there was another emergence meeting. The heads of state gathered together, and the announcement was made concerning this meeting. There were tut's and grunts of disbelief. The head, head speaker said. 'I am Sorry I forgot to mention that this race of Zogerians had two heads. 'I know he said that it is not a long time since we had an emergence a meeting over this problem but it needs seeing to.' The secretary got up, 'if this crops up again Soon then we will be getting paid extra.' There was a cheer, please remember that a cheer on the planet Zog is worth two on Earth. 'To get back to the problem,' the secretary said, 'it seems to be happening on a regular basic it was five billion years ago when this last happened. Has anyone any ideas?' remember that there was two heads to every Zogerians that made the silence twice as quiet as here on Earth. To understand how this effect Zogerians was that they all needed hearing aids that's four per Zogerians. There were six arms raised. I should explain this was each Rogerian had six arms each. 'Yes,' the secretary said, 'we could send Someone down to keep an eye on things. Please note I did not point out that they only had one eye because I said that they would send Someone down to keep an eye on things. That is one eye. To keep things clear they only had one eye between two heads. 'Let's keep things as they are,' the head, of state said. That suggestion was carried unanimously. I thought I would leave the wrong spelling in to show what a beginner I am. I have not included any more mistakes in this book as it would look like it had been attacked by a clutter of spiders that had just walked through a can of spilt paint. As we know there were five ice ages. Maybe on the last ice age they did send Someone down and they are the Society that is promoting the idea that aliens are populating our planet. Now I hope you can see the problem with this conception. This is of course not including the five periods of destructive by the droughts.

If we look at another idea of what happened, and it can be no more or no less fantastic and that is a creator. What have we got to show that this is just as fantastic as the alien invasion? There are no signs of aliens being dug up on any of the architectural digs. The cave drawings were etched on the walls of the caves of the animals and birds that where in his environment. There are no dinosaurs nor is there any form of space ships if there had been there would have been a special cave with these drawings on the wall and there would have been an entrance fee. If you think about it the most natural thing that could have destroyed everything and leave no trace is water. Let's do a quick total of the disasters. Five ice ages, five droughts, two floods. And there were numerous times that meteorites have struck and the world was brought to its knees, by volcanic ash blotting out the sun. Even the physicists and scientist must now realize that the amount of times that the world has been destroyed. The most unrealistic thing is that the Earth's environment as we know it today had to start from the beginning each time. Now the way the time scale has been happening we must be getting near the next disaster. The bible says that God would need to help the Earth from fire and brimstone raining down. As with all the other disasters this is the one that is most likely to happen. The ending of the universes sun's all came about from there cooling off and then bang. It should have taken billions of years for the Earth to reach this point but with global warming it could become Sooner than later. We know that the weather all over the world is going crazy. To have the Earth come to the conditions that fire and brimstone to rain down then it must come from volcanoes and Earth quakes. The scientists have said that the bibles stories about miracle can be explain by natural occurrences. I will point out later why this can be disputed.

Looking at the way things are going let's hope we can get global warming under control.

A thought has just come to me if the Earth is the way it is through evolution, then why over the last fifteen billion years have the worms not evolved they haven't the sense to stop getting up early.

I suppose they realise that if they did then the birds would have a sleep in, there cleverer than a worm.

Let us get back to probabilities of how the Earth keeps managing to recuperate after all these disasters. We have gone through the alien populating the Earth eight times. Now let us look at bacteria coming from out of the sea. We have bacteria on everything that is here on Earth So why would they must go through billions of years of evolution to get on land when we already have bacteria that can breathe. Now we have the same problem they would must start from the beginning, eight times. Let us look at it in time span. The Earth is fifteen billions of years of age. We have had the world reformed eight times that is fifteen billion divided by eight. That is one billion and one. In the late March, 2012, European astronomers announced that they had discovered a planetary system around a *metal-poor* star, that is, a star containing mainly hydrogen and helium, the two ingredients thought to have been present at the Big Bang. In other words, this star and its two Jupiter-sized planets appear to be survivors from the extremely early universe. The star is HIP 11952, and it's not the only very ancient star known to have planets. But, at an estimated age of 12.8 billion years, this exoplanet system is one of the oldest systems known So far.

REPLY

If there is going to be any life on any planet then surly this planet would be the first to be populated. As I have said every galaxy in the universe has the same environments So there is no logical reasoning why the sun's satellites cannot contain what is here on Earth. The more planets that the astrologers find that are older and owing to their environment they could support life, (but they don't.) makes more than likely that there must be a creator.

We know these are not planets like our own Earth. Our sun is at least a second-generation star. How do we know? We know because the sun and Earth and everything around us on Earth, including our own bodies, contain chemical elements heavier (more complex) than hydrogen and helium.

All chemical elements heavier than hydrogen and helium are thought to have formed stars, via the process of thermonuclear fusion that enables stars to shine. These elements or metals were released into space via supernova eruptions. It's the "we are star stuff" idea that Carl Sagan popularized a few decades ago and that still reSonates with So many. Sagan Also said: We are a way for the cosmos to know itself. In the video, Sagan is talking about the universe we know best: our Earth, our own Solar system, other second- or third-generations stars that lie relatively near us in space. The HIP 11952 system is not like these familiar stars and planets. Instead, the system is a relic from a much earlier era of the cosmos. And So astronomers want to use these planets and their star HIP 11952 to begin to understand that early time in the universe, say, 13 billion years ago. For example, we believe we know how planets like our Earth form. We believe they form from vast rotating clouds of gas and dust swirling around a star. Presumably the first planets formed in much the same way, but no one can be sure. Also when did the first planets form? The planetary system HIP 11952 might help lead astronomers along the path of finding answers to these questions. These astronomers found the planets around HIP 11952 while conducting a survey targeting especially metal-poor stars. They say planets around such a star should be extremely rare. Veronica Roccatagliata of University Observatory Munich was principal investigator of the planet survey. She said in a press release:

In 2010 we found the first example of such a metal-poor system, HIP 13044. Back then, we thought it might be a unique case; now, it seems as if there might be more planets around metal-poor stars than expected. Pasquali from the Centre for Astronomy at Heidelberg University, a co-author of the paper, added: We would like to discover and study more planetary systems of this kind. That would allow us to refine our theories of planet formation. The discovery of the planets of HIP 11952 shows that planets have been forming throughout the life of our universe. While conducting a survey of metal-poor or very ancient stars, European astronomers discovered one of the oldest planetary systems known So far. HIP 11952 is now known to have two Jupiter-sized planets. The system is thought to be Some 12.8 billion years old. Astronomers hope to use this system to begin to understand how and when the first planets formed in our universe. Am I missing Something I thought the Big Bang was the reason that we have galaxies and all that they behold? Stéphane Udry: Evidence of billions of rocky, habitable planets in our galaxy

REPLY

All these different experts seem to have different theories about the same subject. I will not give a reply to each individual as my one answer will cover them all. I have already answered these theories before So I don't see the point of answering them all again.

BLACK HOLES

The Milky Way has 100 billion planets, astronomers say

Imagine what it is that it has the power that is So strong that nothing can escape from it.

Not a planet and not even light. That's what black holes are a spot where gravity's pull is huge, ending up being dangerous for anything that accidentally strays by. Now that's what I call getting a grip on Something. But were did black holes come from, and why are they important? Below we have 10 facts about black holes, just a few titbits about these fascinating objects.

I would like to put forward my logics about the ten following facts. The above states that they have ten facts about black holes that is wrong information about black holes they are not facts only theories. My writing on this subject is based on logic and what I have observed in the readings of information given by astrologers, scientists and physicists. But most importance what I think is the right explanations for these observations. These will be answered after every each fact that has been put forward. That seems prevalent.

Fact 1 You can't directly see a black hole

Because a black hole is indeed "black" — no light can escape from it — it's impossible for us to sense the hole directly through our instruments, no matter what kind of electromagnetic radiation you use (light, X-rays, whatever.) The key is to look at the whole's effects on the nearby environment, points out NASA. Say a star happens to get too close to the black hole, for example. The black hole naturally pulls on the star and rips it to shreds. When the matter from the star begins to bleed toward the black hole, it gets faster, gets hotter and glows brightly in X-rays.

REPLY

You can? And you don't need glasses.

They have spent, millions, billions or is it trillions on telescopes and space exploration. Give or take a bob/dollar or two, depending on what currency you use. Let us use what facilities we have here on Earth. This costs no money to try out this experiment. I believe that if you want to find out what is happening in space then logic and the reSources that we have here on Earth can be used. You could build expensive Labourites, or to make sure no light can enter then you could build an expensive bunker underground. If you can convince the government you could apply for a grant. To do this you will have more chance if you got together a committee to fight in your corner. These of course are professionals and will must be paid for their knowledge in government affairs. After fighting your case you then receive this grant for millions of pounds. It's night time. That should save a couple of bob/dollars. This experiment can be done by any house holder that can wait for night-fall. I stipulate they must have a room to enable to do this experiment. I think if they are house holders then they will have a room. If they are not house holders then they will need to apply for a grant. This is on the advice of Solicitors. Who will recommend a committee to fight your case? I could go on and find a lot of reasons to spend more of the tax payer's money. I want to keep it simple. Sorry I seem to have digressed.

THE EXPERIMENT

It's still night time you are in a dark room. To do this experiment you will need Some light, not in the room but from the outside. Here a grant could come in handy. Have a form of light on the outside of the room, say like they use on a sports field. We are nearly there. You are in this dark room with light blazing down.

That's what you could, but with my experiment you have a dark room and I flash light outside. You will need to turn it on, Sorry about the expense. If it is a quiet night with no wind you could use a candle. For those people applying for a grant this will not be liable to be repaid.

You are standing in this dark room what do you see a light on the outskirts of space, the space being the darkness in your room. If you were billions of miles away and you were an alien space ship with these blue aliens on board. The reason they are blue is I don't believe in little green men from outer space. What do you think they will see? They of course would see a light surrounded by blackness, this is what you see. To put it in to prospectus the darkness of your room would extend as space does and go on forever. You now multiply your room a billion times. There it is a black hole. But we mere mortals know different. It is only darkness.

Fact 2: Look out! Our Milky Way has a black hole.

A natural next question is given, how dangerous a black hole is. Is Earth in any imminent danger of getting swallowed? The answer is no the astronomers say, although there is probably a huge super massive black hole is lurking in the middle of our galaxy but Luckily, we're nowhere near this monster we are about two-thirds of the way out from the centre, relative to the rest of our galaxy — but we can certainly observe its effects from afar. For example: the European Space Agency says it's four million times more massive than our Sun, and that it's surrounded by surprisingly hot gas. *Sagittarius A in infrared (red and yellow, from the Hubble Space Telescope) and X-ray (blue, from the Chandra space telescope). Credit: X-ray:NASA/ UMass/Dewan et al., IR:NASA/STScI*

REPLY

So they are trying to scare the pants of me. The astronomers say there probably is a super massive black hole in the Milky Way. My understanding of the astrologer's point of view is that this massive black hole is drawing all matter near it into the black hole. But because we are on the outside of the Milky Way we are in no danger. My thoughts on this are, if asteroids, planets and sun's are travelling at seventeen thousand miles per second, outwardly and as I have said in the previous reply space/black holes do not move. If black holes moved with all the other matter then they would must move in all directions to swallow anything up. The astrologers say this black hole is four million times more massive than our sun, if everything near a black hole is being drawn into the black hole. How the hell can the universe be expanding. Sorry about the swear word but I couldn't control myself. The reason given for this statement is because of their strong gravitational pull. (I will explain later why this is not possible.)

Fact 3 White Dwarfs

They say you have a star that's about 20 times more massive than our Sun. Our Sun is going to end its life quietly; when its nuclear fuel burns out, it will slowly fade into a white dwarf. That's not the case for far more massive stars. When those monsters run out of fuel, gravity will overwhelm the natural pressure the star maintains to keep its shape stable. When the pressure from nuclear reactions collapses, according to the Space Telescope Science Institute, gravity violently overwhelms and collapses the core and other layers are flung into space. This is called a supernova. The remaining core collapses into a singularity — a spot of infinite density and almost no volume. That's another name for a black hole.

REPLY

They are doing it again. My grey cells are being devoured by a black hole I will explain as I go along. Apparently, the centre of the Milky Way has a massive black hole that might engulf us and that's the end of that. But don't worry we are too far away for this to happen. They have just said that our Earth will fade away into a White Dwarf. This is not possible if you go by my logics on the subject on the way gravity works then you know that this is not possible. There is Also a massive sun that is about twenty times bigger than our sun. This is going to be put under that much pressure that it will collapse. This will happen because gravity will be stronger than the sun's way of stabilising itself. The sun's spinning forces keep all its contents in orbit around it. The Boffins say that gravity will be stronger and course it to implode. Gravity does not work like this it has no control over the way the Solar system works. It has no influence on Earth's way of working. Then blow me down gravity takes over and causes it to collapse even further. The iron is the core at the centre of this collapse and blow me down the gravity does a U-turn and blasts everything into space; this includes most of the core. Not all of it there is Some left. Here comes another problem how the physiatrists and scientists know this. With all their technology even, radio waves cannot return to Earth as not even sunlight can escape from a black hole. This is according to our experts. And I thought I was indecisive. What is supposed to happen is when this massive sun explodes it collapses. When since the beginning of our technology has it shown an explosion collapsing inwardly? Let us continue with their theories. This sun implodes because the gravity is So strong. The core gets drawn to the centre of this scenario. This would of course be iron as iron is the heaviest metal. This is now a black hole. Within the centre being a core of iron. Can we go back a little bit they have just said they have received data about a black hole. How is this possible the radio waves could not have returned as a black hole abSorbs everything that enterers its domain cannot escape. Another of my grey cells have just been destroyed.

Fact 4: Black holes come in a range of sizes.

There are at least three types of black holes, NASA says, ranging from relative squeakers to those that dominate a galaxy's centre. Primordial black holes are the smallest kinds, and range in size from one atom's size to a mountain's mass. Stellar black holes, the most common type, are up to 20 times more massive than our own Sun and are likely sprinkled in the dozens within the Milky Way. And then there are the gargantuan ones in the centres of Solar system, called "super massive black holes." They're each more than one million times more massive than the Sun. How these beasts formed is still being examined.

REPLY

*We seem to have three different types of black holes. P*rimordial are the smallest. Stellar black holes are twenty times more massive than our sun. The largest are super massive. These are one million times more massive than our sun. Now that's big. They do not know where these have come from. If you want to understand the working of the universe then by following my observation then there is nothing unusual in these monsters being in the universe

Fact 5: Weird time stuff happens around black holes.

This is best illustrated by one-perSon (call them Unlucky) falling into a black hole while another perSon (call them Lucky) watches. From Lucky's perspective, Unlucky time clock appears to be ticking slower and slower. This is in accordance with Einstein's theory of general relativity, which (simply put) says that time is affected by how fast you go, when you're at extreme speeds close to light. The black hole warps time and space So much that Unlucky time appears to be running slower. From Unlucky perspective, however, their clock is running normally and Lucky's is running fast.

REPLY

Einstein has got it wrong. Time is not effected by the speed a perSon is travelling at. Lucky is watching Unlucky as he falls into a black hole it appears from Lucky's position that Unlucky has speeded up because of the gravitational pull of the black hole. Let's fetch things down to Earth. A comet passing through space is travelling at seventeen thousand miles per second. It enters the Earth's atmosphere does its speed accelerate. Or maybe it slows down. I do think So. It is not possible for a black hole to warp space or time. Let us just take a look at this from Earth's prospective and a simple experiment. If you are on an aeroplane does your watch go faster when the plane goes faster. Don't get me wrong there are circumstances when your watch does not keep up with time my watch had lost an hour when I got off the plane at the Paris airport.

Fact 6: The first black hole
   This wasn't discovered until X-ray astronomy was used.

*Cygnus X-1 was first found during balloon flights in the 1960s, but wasn't identified as a black hole for about a*nother decade. According to NASA, the black hole is 10 times more massive to the Sun. Nearby is a blue super giant star that is about 20 times more massive than the Sun, which is bleeding due to the black hole and creating X-ray emissions. *Illustration of Cygnus X-1, another stellar-mass black hole located 6070 ly away.*

REPLY

I hope there is nothing in the pipe line for building Something stronger than our X-ray astronomy

Cygnus X-1 is a black hole. Nearby is a super star that is twenty times more massive than our sun. Apparently this sun is being sucked dry by this black hole. Is there nothing safe from these black demons? And this is creating X-RAY emission. If there is a sun then it must be as our sun and giving off gases. NASA/*CXC/M*. Weiss Fact 7: The nearest black hole is likely not 1,600 light-years away. An erroneous measurement of V4641 Sagitarii led to a slew of news reports a few years back saying that the nearest black hole to Earth is astoundingly close, just 1,600 light-years away. Not close enough to be considered dangerous, but way closer than thought. Further research, however, shows that the black hole is likely further away than that. Looking at the rotation of its companion star, among other factors yielded a 2014 result of more than 20,000 light years.

REPLY

So the nearest black hole to Earth is estimated to be 1,600 miles away. Knowing what these beasts can do I think I will move to Somewhere safer? Going by the information above I see that they are doing it again, talking about hedging your bets a scientist would is proud of NASA/CXC/M. they could probably get a job in the banking sector. For those who cannot see these bets being hedged are, a black hole draws everything into its interior, and then they say there is no danger from our black hole. Maybe ours is not as black as it looks. The universe is fifteen billion years old and black holes must be the same age. If black holes are drawing everything in how is there anything left in the universe.

Fact 8: We aren't sure if wormholes exist.
A popular science-fiction topic concerns what happens if Somebody falls into a black hole. Some people believe that there is a worm hole that takes you to other parts of the Universe, making faster-than-light travel possible. But as this SmithSonian Magazine article points out, anything

is possible since we still have a lot to figure out about physics. "Since we do not yet have a theory that reliably unifies general relativity with quantum mechanics, we do not know of the entire zoo of possible space time structures that could accommodate wormholes," said Abi Loeb, who is with the Harvard-SmithSonian Centre for Astrophysics.

REPLY

I am lost for words. The physiatrists do not have a theory. I am flabbergasted. Come on my reader if you can think of a theory then get in touch with a university I am sure they would be very pleased that they have been given a Solution to their problem. Come on you brains of our Society. How can a worm hole exist when nothing can escape from a black hole?

Fact 9: Black holes are only dangerous if you get too close

Its okay to observe a black hole if you stay away from its gravity pull before it starts grabbing you. Think of it like the gravitational field of a planet. This zone is the point of no return, when you're too close for any hope of rescue. But you can safely observe the black hole from outside of this arena. By extension, this means it's likely impossible for a black hole to swallow up everything in the Universe (barring Some Sort of major revision to physics or understanding of our Cosmos, of course.)

REPLY

If what they say above is true then surly it must apply to other objects in the universe. What I am referring to is the statement that when an object enters a planets gravitational field it stays there. This applies to asteroids as this is the only matter that enters Earth's atmosphere. Where have all the comets escaped from. I thought I had read about black holes gobbling everything up.

Fact 10: Black holes are used *all the time* in science fiction, So are worm holes

There are So many films and movies using black holes, for example, that it's impossible to list them all. Interstellar journeys through the universe includes, a close-up look at a black hole. Event Horizon explores the phenomenon of artificial black holes — Something that is Also discussed in the Star Trek universe. Black holes are Also talked about in Battle star: Galactic, Star gate: SG1 and many other space films.

We are on the last of the ten facts, concerning black holes; there have been many theories on this subject. I hope Someone will take my observations serious. These I will tell you what they are at a later date. We have had a lot of talk concerning the universe from the physics point of view. I will now look at the other side of the argument. For all none believers don't switch off as I hope I can keep you attention with Some funny tales. Two birds were flying over a mountain one sees, the ark on top this mountain.

One says to the other, 'I've just seen an ark on top of this mountain. I don't think that bloke was expecting a flood,'

The second bird said, 'I have had a good look around and considering all the water it's going to take a hell of a drought to get rid of this lot. Also if you note on where they have decided to park they are in dead trouble.'

'On what basic do work that on.'

As you can see they are parked up on top of Everton Brow and you know what a sticker Liverpool council is against illegal parking The Lord said to Noah, 'I want you to build an ark as it will rain for forty days and forty nights.'

'If you're sending us to Manchester Noah replied I will need an umbrella.'

Noah asked god, I had a dream that all my long hair was cut off, and I had lost all my strength, does this mean I am SamSon.

'No.' said God, 'if you had all your hair cut off you would still look like Delilah

ICE FOUND ON ASTEROIDS

We must look at this from the beginning. An asteroid is a chunk of rock. These are travelling through space at fifteen thousand miles per second. We know that space is at absolutely zero. We Also have the sun's rays. We know that the sun is hot. We Also know that the sun's rays have no effect on the way space reacts to it. It does not heat up space. What we do have is three different components, the sun's heat and the absolutely zero temperature of space coming together, lastly the asteroid this is a chunk of rock that is heating up Come on my reader we have learned a little bit about space what do you think will happen. The asteroid is travelling through space as can be seen by the trailing stream steam the reason for this is as the asteroid travelled through absolute zero temperature it collects ice. As the sun's rays hit the asteroid it courses the ice to evaporate. Therefore, the physiatrists and scientists have found ice on asteroids. So instead of spending trillions of dollars on space probes I would put this as a dollar or two less. There is no water in space water can be found around sun's/planets and gases are around planets that have been caught in their penny force. If space contained huge amounts of water, then surly you would get a deflection from the sun's rays. And this would create either a glare So bright that you can forget about black holes or there would be a terrific show of rainbows. This water would then turn to ice this means that all space would contain meteors and these will be big.

When a planet is So far from the sun, the sun's rays are not strong enough to melt the ice. You can see this in a planet that is not a gas planet. What we must do is look at a planet that is nearer to our sun and is a rocky planet. Mars is the one we need to look at.

This Also applies to gas giants. As the planets are travelling in an outward direction they will be further away So the heat will be of a lesser strength. As asteroids contain h02 this forms into ice. Let us now get a grip of the ice age scenario. Please wear gloves. As it has been mentioned there has been at least five ice ages. I went outside yesterday and I thought it was the start of another one. I will explain later why there never were any ice ages. As my explanation will show it Also covers why tides do no go in and out. I will be waiting for the physitrists, scientists and the geologists to tell me to get down to the beach more often. And I expect that when I first get there they hope the tide is out.

What could it possibly that prevents an ice age? The answer is below.
Why the moons gravity field not affect the Earth's tides?

I have looked up on the internet to find out what effect the moons gravitational force has on the Earth's tides. If you are a beginner like me, forget it. There is that much jargon on this subject that my mind was spinning faster than a whirlwind in a hangover. The experts that have studied space will know this but if they look it up to be certain they have my blessing. The explanation goes on for about six pages, don't quote me on this. As I have said about six pages and I understood about two paragraphs, Sorry parts of the two paragraphs. So missing out the parts I don't understand as I do not want to show my ignorance on this subject I will pick out the main parts that I disagree with here goes fingers crossed. Gravity like the sun's rays affects the Earth's surface in the same way it does not matter if you are in Timbuctoo or Milton Keyes on a bad day. You have the moon in one constant position all of the time in reference to the Earth's position. All my readers know this; it is a well-known fact that we only see one side of the moon.

The gravitational force coming from the moon is affecting the Earth's entire surface at the same strength. So tell me how it can move the tides. To do this its gravity would must surround the Earth and then does a ninety degree turn. There is a rumour going around spread by physiatrists that the sun's gravity influences the Earth's tides. Phooey. The sun's gravity has no effect on the Earth's tides the sun controls our Solar system. The Earth's spin controls our penny gravity. This holds matter in our atmosphere. If you think about it, I wonder what happens when the sun's gravity is pulling one way and the moon's gravity is pulling the other way this is when they are on opposite sides of the Earth. I know one thing I am not going anywhere near the beach.

Gravitation is caused by the planets spinning. The reason that tides change is because of the Earth's tilt. This means the Earth tilts, and the water stays in the same place. That gives the impression that the sea is moving up the beach. And that it is moving out from the beach when the tide is going out. Let us do an experiment, I would like to ease my reader's fears that there will be no harm done to one's perSon, but they can then say that they have been involved in a scientific experiment. Take a bucket if you don't have one borrow one. Fill it half full, this does not must be precise as any amount of water will do if you are environmentally friendly just enough to cover the bottom of the bucket. This will help if you are not at your peak condition, you could always get your granddaughter to help. Why well you must lift it off the floor. Now tilt it. You might think that being a scientific experiment you need to know which way to tilt if. I will leave that to the individual to make up his own mind. Look at the result very carefully. The bucket will move one side will move up the other side will move, wait for it. Down. If this does not happen this is either you are not doing it right, or the water is not liquid enough or the bucket has a hole in it. When you have Sorted this out, try again. Ready steady go. What happens? The water does not move the bucket does. The bucket represents the Earth. The sun does not have any effect on objects, whether in space or here on Earth.

We need to look at how cold reacts to heat. I will jog your memories. Earth, that is our best way of looking at problems and as they all seem to be acting differently.

THE EARTH

Our Earth is travelling through space as do all other objects. But it reacts differently to the absolutely zero temperature. This is because of our ozone layer. Next let us take a quick look at an asteroid. With the two separate degrees of heat and cold in space we get evaporation. This you can see as a trail behind the asteroid. This asteroid is belting along like a bat out of hell. There is nothing else only space it then comes into a sun's rays now we get a transformation. You have heat and cold in the same place, what happens. You get condensation. Now the strongest of the two is the absolutely zero So the condensation freezes. That is why an asteroid has ice on it. But as it gets nearer to a sun the ice will melt. This is why you get the stream of steam following it.

Now we have the moon. It seems that the larger objects such as planets and moons do not react in the same way as smaller objects that are asteroids. This is not such a tremendous discovery as you might expect it to be, why. When the Big Bang occurred, and all the matter was blown into the universe. Then the crust and the plasma, and gases that were hurled into space travelled through space that was at absolutely zero. These objects whether being a Solid or of a molten composition would start to feel the cold (who wouldn't just thinking about that makes a shudder goes down my spine.) the molten plasma (a sun) would gradually cool down and in time would obtain a crust. After a while the sun's crust would get thicker and thicker. All the matter in space would all be affected in their different stages. The first to go through this change would be the sun; this is because the sun was the first object that was around at the time thus the Big Bang. Everything else in the universe came from the sun, Some in planet form and Some as gases others as chunks of rocks, asteroids, Also the gases that are in our universe are not affected as sun's and moons are.

Gases are a different aspect from other matter as it reacts differently to how it is affected by the universes way of controlling things. If you think about it Earth is on the last stage of these changes. if So bang goes the Earth? If you think about it when a new galaxy is born it is not by a planet exploding but the sun in that galaxy. So does our Earth finish up as a moon or just a dead piece of junk in the universe? As we are the youngest galaxy then we would have seen this in the Solar systems from the past. Let us look at the scenario if a planet did explode what do you think would have happened. The first thing that it would shatter, and all the Earth's' contents would be blasted into our galaxy. The bible says that there will come a time when the earth will be rained on with fire and brimstone. Do other moons have the same properties as our moon? This in effect could explain where comets have come from. It does seem amazing that with the amount of matter that the first Big Bang accrued there is not a lot more planets and moons. The only explanation I can think of is there was a thick crust that would mean that the chunks were a lot bigger. The problem is that the first bang created all the Milky Way. This seems reasonable as that would explain a creator. How have I come to this decision? Well if we look at the Milky Way and the way it has evolved we noticed that it consists of a sphere this has a uniform pattern and there is no variation in this all the way around its circumference. Then all of a sudden there appears another galaxy that is not in the sphere. It sticks out like a wart on a nose. I was going to say like a pimple on a bum. The reason I have made this compariSon was that the most likely perSon to see a pimple on a bum is a doctor. (Unless you are very friendly with your milk man:) I thought it would be better to inform my other reader. I have not seen any reference to this only that they are saying that our galaxy is the second generation of Solar system. In the circumference of the Milky Way this has only happened once. Maybe the physitrists or the scientists can explain this as it all goes beyond the case for us beginners. Here we go again this idiot believes in a creator. The universe is fifteen billion years old and as we can see evolution does take place I am not disputing this I suppose if the truth be known when god create life it is common sense to us now a days that this is what happened. This recalls God probably said go forth and multiply. Then again, he wouldn't need to as he would have known this. What this is leading to is how come man's evolution has only accrued over the last five thousand years. And the other evolutions has taken billions of years. I wonder why Charles Darwin never thought or mentioned this. Let's have another look at the Milky Way and its age. What does this tell us about what there is in the future? The time it has taken from the Big Bang to our universe's existence. Our universe is fifteen billion years old, and the Earth is the second generation. Then there must be other sun's in the Milky Way ready to blow their selves to bits. Wrong. How have I come to this decision? Observations. Based on my understanding of observations on how the universe is working then these cannot be wrong. We know by our observations of the universe that sun must reach he stage were the absolutely zero temperature has cooled the sun down to the stage were the crust has got to the thickness were the heat being contained by this procedure it then must explode. If we look at the Milky Way can we see if any other sun has reached this stage? If So which sun exploded and where was it in the Milky Way. Can any of physicists tell me this simple physic problem? If none of the sun's are anywhere near the stage of blowing its top, where oh where did Earth come from if the universe is fifteen billion years old and there is no sign of another sun disintegrating. All the universe was created at the same time why is our galaxy the only one from billions of galaxies where this has happened. What would happen if a sun in one of the galaxies exploded? The sun in this galaxy already has a moon or moons and an asteroid in its orbit. These are a long way from its mother sun So the following satellites and all other matter would mean the three would be a lot more moons etcetera. If I can recall I think that the physiatrists could not work out why Some moons were spinning in the opposite direction. I reckon it must be billions of years for this to happen, this should by my logic this galaxy would be a lot larger than other galaxies. This means that the probabilities of there not being a creator are getting to look more probable. I reckon that the physicists are going to find it hard to dispute these observations of mine. I wonder at what stage is there going to be a sun exploding in our universe. Owing to the different sizes of the sun's we would expect them to explode at different times. Let us look at this scenario and think what the consequences would be. There would be different consequences to what happened at the time of the Big Bang. How's this for a logic way of thinking about what is going to happen with the moons and planets.

Let us have a look below at what the scientists say about neutron stars.

Neutron stars are created when a star around eight to ten times the mass of our Sun runs out of fuel. The outward pressure generated by fusion reduces rapidly, allowing gravity to pull the star in on itself and trigger a supernova, where the outer layers of a star's crust gets blown into space. I do not know anything about neutron stars, this means I must look at what I believe in and my observations of the universe. Why they would expect a sun to react differently to the original Big Bang I do not know. You have all these different objects in the universe that are in the same environment. Their environment in space is at a temperature that is at absolutely zero, and that does not change. Why they theorise that the sun is going to react in two opposite ways when in the same environment I do not know. You might be able to get a substance to react differently but that is by changing their environment. What I mean by this is taking a substance and experimenting with it. This means in fact you are changing its environment. So as far a star suddenly decides to react in the opposite way to other stars this is beyond belief. This is not how the universe works. The reason why the sun's components would not collapse into its self is because of Isaac Newton's law on forces. The force of the explosion will send the crust and any other matter on an outward direction. This is because the sun's inner contents weigh more that the outer layers. Therefore the contents would all head in an outward direction as this is the only place where there's no resistance. The blast from all sides would be equal at the centre of the sun. This means they would not affect the centre core or any other matter that does not interfere with the weight ratio. Nothing gets pulled into itself, this is explained later. As you will note by my observations of the universe and it brings into the equation Isaac Newton's law of forces. I am glad I have Someone on my side. Sun's do not run out of fuel. You might ask how the hell has he worked that out. There is an easy explanation and if you have been listening you will know that if the sun's started losing their power to produce more fuel this means less heat, then my explanation of a sun cooling down means this in fact would mean that the sun would react quicker to its environment of absolutely zero thus it would cool down quicker So emphasising my observation that no way could it burn itself out and turn into itself. They have the above right concerning the outer layers getting blown into space. Just what do the outer layers consist of? Look at the earth's surface. This is an earlier version of the sun starting to cool down. Let us take it a bit further. The crust would increase until it came to the point that it would be like a pressure cooker with no release valve. Remember there would not be any water content. To my understanding the spaces that surround the sun consist of gases. As can be seen by our sun there are all the colours of the rainbow. The sun is molten plasma. Why would this liquid, and its core stay. Let us look at what Isaac Newton says about forces and how they work. You have equal forces pushing against each other. They then stay in the position where they meet. This applies in the situation when there is no escape route. Now I have Sorted all that out for the physicists what can I do next. Let us look at an atom that has been produced from a collapsing star. The remaining matter continues to collapse under gravity, forcing electrons and protons to be squashed together and become neutrons. The neutron star will have less mass than its parent star (typically about 1.4-times the mass of the Sun), but this mass will be confined by gravity to a region of approximately 20 kilometres (12 miles) across, leading to an incredibly dense object. It is this density (a teaspoon full of neutron star would have a mass of about a billion tons) that truly defines a neutron star. Come on which neutron planet are you on, let's not go overboard I would like to know how many people on planet Earth believes that. As I have not heard any contradictions on this subject then I must be the only one that doesn't believe it. The reason why is because of my theories, Sorry my understanding of how the universe works in connection to Earth's physics. I hope that explains why this theory has been crushed into a size smaller than an atom, and weighs nothing. Let us see what the physicists say about a star imploding. They say when a massive star explodes, not all the material is ejected into space. Some of it collapses into an extremely compact object known as a neutron star, inside which gravitational forces crush protons and electrons together, turning them into particles known as neutrons. when sa spoonful of Come on does any-one think that this is possible. I am no physicists, nor am I the thickest perSon on this planet, (this is only my opinion!) these are physicists that are putting forward these theories. irirA metal will not crease in no matter how much pressure it is put under. Once all the air has been dispersed theWhat can be the best way to illustrate this? Think! How about Something easy. Putty or water this seems like a good way of showing this experiment. If you put this in a sealed container The As far as a water experiment goes let us look at the ocean. As you go further down, the pressure increases. What is it that increases the pressure? Is it the weight of the water above? How can this be there is plenty of room for the pressure to escape to. We know there is less oxygen. This brings us to the answered question once you have dispersed all the oxygen you cannot increase its weight. Let us take a container down with us. When we arrive at the lowest point that man can stand we place a lid on the container So sealing the water inside. We go back to the surface and weigh the bucket. Do you think that the container will weigh heavier that a container of the same size with water taken from the surface. Of course it won't. How about this, the only difference to this environment is the human body. The human body will feel the extra pressure because it is the least capable of standing the pressure. Plus the oxygen in the human body disperses that is why a human being gets crushed to death you might ask what about a submarine. "Think" okay, got it the submarine goes down deeper that mean's the less oxygen, So the higher the pressure. The submarine contains lots of oxygen that means it collapses. Have you noticed when you have seen a submarine in a nature study film and they are at a great depth you will see fishes swimming about. They are more that likely be described as Deep sea creatures. The following is from Wikipedia, the free encyclopedia. The Common fang tooths. *Anoplogaster cornuta*

The term deep sea creature refers to organisms that live below the photic zone of the ocean. These creatures must survive in extremely harsh conditions, such as hundreds of bars of pressure, small amounts of oxygen, very little food, no sunlight, and constant, extreme cold. Most creatures must depend on food floating down from above.

These creatures live in very demanding environments, such as the abyssal or hadal zones, which, being thousands of meters below the surface, are almost completely devoid of light. The water is between 3 and 10 degrees Celsius and has low oxygen levels. Due to the depth, the pressure is between 20 and 1,000 bars. Creatures that live hundreds or even thousands of meters deep in the ocean have adapted to the high pressure, lack of light, and other factors.

GRAVITY/WEIGHT

Gravity does not hold things down; the physicists us gravity and explain it as a force that pulls objects to the earths surface.

WRONG

It is the weight of the object that has the control of the matter not gravity as defined by the physiatrists. So it does not increase their weight. If gravity had any effect on the weight of objects, then surly the heavier object would fall faster. What would happen if every object fell to Earth judging on its weight? The heavier objects would pass the lighter ones. This would mean that you would have more objects floating about in our atmosphere. I can just imagine this situation. We would be dodging these floating obstacles. And as they got lighter they would be caught in the wind. Can you imagine a floating mass of feathers travelling at ninety miles an hour? This really would give the true meaning when you say you could have knocked me down with a feather. Let us take this a bit further as in lighter or heavier depending which way you want to look at it. Let's say one hundred weights. It certainly would make life interesting it these weights where flying about. You might as well ask how it could be possible for this to happen. There must be many times when heavy objects have dropped from a high rise construction site. Now you can see why every object must fall at the same rate. The above was just me passing the time but I hope this made you think about gravity.

Gravity takes place by the spinning of a planet or a star. This is the physiatrists view. So where has the gravity from the collapsing of neutrons. This is reacting to the exact opposite every we have come to believe.

Now by the reckoning of the above facts a black hole is created. Later you will come across my logic of black holes.

Let's go deep down to see if we can find the explanation in our deep blue seas. As you go deeper into the depths of the sea the pressure increases. At these depths the pressure is that great it can crush a submarines hull. Now I call that pressure. If the pressure can crush a submarine how come there are fish swimming at this depth? This means that you can make a liquid heavier. This by my way of thinking shows that there is less oxygen the deeper you go down. As I have pointed out there must be an exit route. largest escape route in the world and that is the sea itself. What is the difference between molten plasma and water? The difference is in the thickness, but they will both react the same way under pressure. Not the same as the Everest's theory about the planet fitting on a table spoon. The way the physicists work things out in the universe is *everything* is nothing like it is here on Earth. Everything is bigger, stronger, brighter, faster, and heavier, and anything else that you can think about. Going by this scenario then the spoon must be big, Sorry we are talking about space, and it must be tremendous. Let us take a step in another direction. You have mountains were the highest one is Everest at 8.529 feet. That's a lot of rock. I hope you can see where I am going with this. Well just to prove you are right I will confirm it within the next sentence. Get a pick and hack out a chunk from the bottom of Everest. (I would get permission first just in case it affects the height of the mountain. especially if it is the highest mountain in the world.) let us go by the physicists reckoning you will need a crane to pick it up. This is another fact that needs looking into and that is that Everest is in fact not the tallest mountain, there is taller

http://www.independent.co.uk/news/science/mount-everest-isn-t-actually-the-tallest-mountain-ecuador-s-chimborazo-beats-it-a7034656.html mountain it is actually Chimborazo - a stratovolcano in Ecuador that's part of the Andes mountain range - because it's the furthest point from Earth's centre and, therefore, the highest in terms of distance. This is 6,248 metres above seas level. Everest is 2,600 metres above sea level. Chimborazo is 3,659 metres from the Earth's centre. Mount Aconcagua, which rises 22,828 feet (6,961 metres) above sea level is in Ecuador it's a stratovolcano, it is part of the Andes mountain range. This is taller than but in reality; it isn't - not according to science. Yup, according to Eli

Rosenberg for The New York Times, Chimborazo's summit rises 20,500 feet (6,248 metres) above sea level, which is shorter than Everest by 8,529 feet (2,600 metres), but that all changes when measured from the centre of Earth. The distance increases to 3,965 basically, since Earth isn't flat (I have come to the decision that the physicists have a contradiction when it comes to the pressure put on to water), it bulges outward at the equator and flattens near the poles. This means that mountains near the equator are technically higher than those in other areas, and it just So happens that Chimborazo is almost smack-dab on our planet's waistline, while Everest is 28 degrees north.

Mount Chimborazo. David Torres Costales/WikiCommons

So how much higher is it? Well, according to one report, Everest stretches a distance of 3,965 miles (6,382 kilometres) from Earth's centre. Meanwhile, Chimborazo stretches 3,967 miles (6,384 kilometres). Though it's only a 2-mile (3.2 km) difference, it means everything when it comes to crowning height titles. In fact, those 2 miles are enough to put Chimborazo at number one, and kick Everest out of the top 20. This isn't exactly news, though - NPR ran a report about Chimborazo back in 2007. So why does Everest continue to get all the love, while Chimborazo goes relatively unnoticed? Well, it all comes down to how hard the climb is. If you're a mountain climber, you want the hardest challenge, which is what Everest offers. It takes 10 days to merely make it to Everest's base camp, six weeks to acclimatise, and then the arduous nine-day climb to the top. On the other hand, Chimborazo takes about two days to climb after acclimatising (about two weeks), reports Rosenberg. Also it's important to mention again that Everest still takes the cake when measured at sea level. If you're using that as a metric, Chimborazo wouldn't even rank as the tallest peak. The above information is from Wikipedia. So if you've already made plans to climb Everest and earn your name a place alongside Sir Edmund Hillary's, fear not, because you are still climbing the tallest mountain in the world (if sea level is your metric). After that, you might as well climb up Chimborazo, because that climb will seem like a walk in the park.

The astrologers have now come up with the theory that two stars colliding create a new galaxy. Gases exploding do not create Solids only a different type of gas. The sun is constantly creating gases and this can be seen by the different colours that surround the outer circumference. If our sun is not capable of burning these gases, what has caused the gases to explode in space? How can two stars collide when each star has its own penny gravity field? You will find out later why I do not think that this possible. And let us not forget that the universe is expending. This gives two reasons why two stars cannot collide. Apparently, the physicists say that the universe will all be swallowed up by these black holes. On the other side of the argument they have the opposite theory that the universe is expanding. You will find my way of thinking logically about the way the universe is going to be in the future.

My head is spinning around with all the contradictions given by scientists and astrologers, and all the other bodies dabbling in our universes existence.

All the scientists and any one that has studied the universe all agree on the Big Bang. They have Also agreed to it there was only one sun. If there was Something in space that created the original sun how come it only managed one? So if the Lord created the sun, or we go by Charles Darwin's theories there is only one of them that can be right. I am not disputing his finding on evolution as living things can evolve, but I am disputing that we originated from bacteria. Over fifteen billion years ago the sun blew up. It took billions longer before there were any significance changes to the universe. In fact, thinking about it there hasn't been any changers that can be noticed. You would have thought if anything was going to evolve why it has taken So long. The reasons are that the changes the universe takes So long that we are not able to see the differences. The only change in the universe that we can observe are on the Earth, why is this So. Nothing in all the billion years since the Big Bang has changed. So how is it, that suddenly there appeared a galaxy with the Earth and its living environment in such a short time. What could have caused our galaxy that was So different to other Solar system? The rest of the universe is still there. In the billions of years that Earth has been around with So many of the miracle changes the rest of the universe has not evolved one iota. Thinking about this scenario and why have the animal and insect and all other living things evolved from the first stages of evolution over the last fifteen billion years means that we should have evolved to a greater plain than where we are now. The universe has been the same over billions of years and it is still the same now. Life could have evolved at any time on any of the centillion of planets in the universe So why has is suddenly happened over the last 5,000 years or there about. I am now going to jump to another part of space intriguing facts. It is the moon. The moon photos show signs of where a volcano existed billions of years ago. Now you all know that the physicists and opinion of volcanoes on the moon. If you look closely at these photographs you will notice as I have noticed that they do not represent a volcanic eruption. If you take a look at the moon's surface it appears that on the surface there are traces of volcanic activity. An asteroid when hitting the moon's surface would give the same impression. You will come upon one that is So similar it looks like a copy of an asteroid. You might ask a question of me with this observation, if this was true then why doesn't it look like a crater. As I have pointed out they have a smoother rim. As the moon has no atmosphere and it is not blessed with our ozone layer then it would not be reduced in size. The Earth and the moon where created at the same time. This is a fact they are part of our galaxy. The sun being the star that controls all that is in a galaxy. The effect of the sun on the Earth and the moon should have left them in the same condition. There is Something else you should note. Going by how the universe works then our moon would appear to be in the later stages of its life. This is going by observation that a crust appears on a sun before it comes to the time when it is So thick it explodes. The moon surface is covered with a crust and there is no sign of any place on the surface were all the heat that has been bottled up can escape. So why hasn't it exploded? The most likely reason is owing to it being smaller than the Earth then the plasma contents would a lot less. As the oxygen inside the moon was being used up and not being replaced it would cool down slowly. The time would come when it would all be one Solid crust. So brings to mind where the hell did it come from. You might notice that I use the phrase, hell a lot, this is my swear word you can put in your own. Our moon must be the only satellite of the universe that does not come under the way the universe works. I am Sorry the Earth is the most likely one that does fit in to how the universe works. Let's take a quick look at the other planets and their moons. They say there are planets that consist of gases with their moons orbiting them. These planets, Jupiter, Saturn, Uranus, and Neptune can be seen to be in the shape of a sphere. Here is my point in all that the bible says. Some volcanoes that were created from hotspots would remain in the natural state others would cool down and later they would be sealed over. When I say sealed over what in fact happens will be revealed at a later date. I know I keep referring to a later date, but you would not want to know the ending of a story in the beginning I mean what would be the point; it would be cutting a long story short. Our earth has So many mountain ranges. Why do the moon's surface not these mountain ranges? The Earth surface has had to fight against the force of hurricane winds and rain storms that were a lot fiercer than they are in today's times. That should have made the Earth's surface a lot smoother. They were both subject to the same conditions. What in the world of science has made So much difference to their contour? I wonder why the space explorers think there has been volcanic activity on the moon. If there had been then there would be a small sign Somewhere on the moon's surface. Lets go back a while when there was a space race, the Americans and the Russians were going all out to be the first in space. We all know that the Russians managed to get there first. Sorry a dog was the first So that put the Americans out of that race. Never mind they got the bronze medal for that. The way to show the Russians that the Americans were not going to beat was to land on the moon. This would be a giant leap for man. They did this with one step. After completing this mission what could be the next mission. Two steps on the moon. This is to show what has been achieved after this tremendous step. So what we get is trillions of pound/ dollars spent. Hang on we did get Some moon rock. Now I come to think of it but hasn't geologists found moon rock here on Earth. This has been delivered free of charge by comets. Now we know that the Russians and the Americans have given up anything concerning the moon. What we have now is the news media need to find a story to grab the head-lines. So along they come with a real blinder. The moon landing was a great hoax. This has been going on for So long. With the media, and NASA throwing accusations at each other it seems to be going longer than the Spanish inquisition. The scientists have said that in fact the moon landing was in fact true. Have you noticed who has kept in the back ground and not give one peep? Now if there was going to be anything said about the moon landing this interested party would have jumped on the band wagon. Does anyone honestly think that with all the knowledge that the Russians had at that time they would have sat on their backsides and not said a word?

Their space exploration were as just advanced as the Americans. Do you not think that they would have said Something about the Americans moon landing, if it was a hoax? The craters on the moon shows that the circumference is very large were as a volcanoes centre is a lot smaller this is because one is a direct impact the other is a force from a different direction as an eruption. If you look at the Earth's surface and compare it with the moons why are they So different? There are surfaces on the Earth that have not changed over millions of years So why do these places not show the impact where meteorites have hit it. On the moon there are So many craters you have a readymade golf course just fill the craters in with sand. I can see that the surface has not been affected by volcanoes. In the bible, they were both created at the same time. The Earth's plasma and the crust were brought together at the same time. The moon was put together with mostly Solid material that is why there is no sign of volcanic craters.

There should be a small amount of volcanic action in my opinion there are no signs of volcanic eruptions. Being only one quarter the size of the Earth is would have cooled a lot faster. This in fact means that going by what has happened in the past. Our moon seems unique in the fact that it should not be here. The physicists, scientists, and geologists all must agree on what one of these learned gentlemen say then it must be right. What am I getting to is the word has got around that the moon had volcanoes, and there are signs that show this to be true? My head is going to blow up, but this will not happen until after the moon blows up. Why do I say this? The moon is at a stage that it should not be at. There are other facts that I am not going to dispute, I bet that comes as a change and to others a shock. If there are dust clouds emanating from parts of the edges of where there are signs of volcanic activity. You know my opinion that they are meteorite impacts. I will now tell you why I have come to this decision. If the moon is covered by a deep crust, and these dust clouds are being pushed by an inside pressure the dust must be in fact a very thin layer, So below the dust we have an escape valve. This must be a narrow escape valve otherwise there would be signs of what is causing this pressure. Wait a second, as is my wont I will take this scenario to our beloved planet Earth. We have places here on Earth where the molten plasma shows signs above the ground. I started looking at the way plasma works and after thing about it for ten seconds what I could use that happens here on our planet. It is of course are seaside's. As we all know where there is no rock face then what we have is our beaches. What we need to do is turn this scenario upside down. The sea now becomes the plasma the beaches are of course the land. Let us jump ahead a couple of billion years. This means that the plasma would have cooled down to the state were there was not as much beach. Or even further when there was only a small pressure release escape. Would there just be a few puffs of dust. Of course, not the remaining escape valve would have closed owing to the absolutely zero temperature. This brings me to the point that the pressure from below is So small on the moon owing to the small amount of molten plasma. We now must change the sea water into plasma. I am talking about when there only a medium release valve. What do you think would be the effect this would have on the moon's surface you would then get a depression? There are only two depressions that you can get and that is a volcanic depression or a comets depression. Which one do you think it would be? If it was a sign of a volcanic activity, then the moon's surface would have a lot more mountains. This means a sign of more mountains. I have discussed the Earth and the moon and there relation- ship to our universe. I will point out Something that is So outstanding that I cannot understand how the physicists, geologists and old Uncle Tom Cobbly and all, have missed it. The Earth and the moon are part of the second generation of a Solar system. How is it that none of the galaxies that were produced by the Big Bang are not in the later stages of their life? This has only one explanation to me and that is the original sun had to be around for a lot longer than fifteen billion years, and I mean a lot longer. All the galaxies in the Milky Way were produced at the same time that was when the Big Bang occurred. How is it possible that the earth has reached to the stage that it is in. what I mean by this is the sun's do not seem to be that much difference. Surly the sun's in the other galaxies should be nearer the cooling down stage than we are. Let us have the scientists, and physitrists answer this unusual occurrence. If this was not So then the Solar system would be at different stages. The only galaxy that is different is ours it should not be here. How can that be if we are the second generation we should be at an earlier stage in our evolution than any of the others? There is nothing that the physicists or geologists can explain of how things usually work yet are not working in the case of our Earth and moon. Sorry they do go by how nature and the universe work but not in the right sequence. I have seen images of the moon and the top dogs in space exploration say that there are signs of volcanic eruptions. What other explanations could there be. How about this for an explanation? When the moon was formed/created it was most likely to have Solids and a small amount of plasma in its contents. Some of this plasma would be on the surface but there would be more Solid material. This plasma in fact would give the impression of a volcanic nature. As we all know that all the matter in the universe was the cause of the Big Bang, So all these planets and stars must contain the same substance, maybe in different amounts. And they would Also have produced the gases, these being different colours depending on the type of metals that the planets or sun's had in there make-up, except our Solar system which is the last one to be created.

There is no other way that all other experts in their fields have decided on. This being a known fact and I am going by what they have said about the age of the universe. To me that means that there was nothing. As everybody knows you cannot produce anything from nothing. The Earth and all the matter in the universe could not have evolved. The universe is fifteen billion years old. This is what the scientists, physiatrists and all the experts have been saying. Before the fifteen billion years there was nothing. And then from nowhere a massive sun appeared that can mean only one thing. There had to be a creator. As I have just said you cannot get Something from nothing. This being the case then Isaac Newton's case therefore evolution cannot apply. Once there are forms of life then you will get evolution. There is one thing for sure that is before there was anything in space it was just a void. There can be only one other explanation that is a creator. This being the case then our creator must have been here before the universe as he created the universe. When and how this miracle occurred, no one will ever know. Einstein said that he wanted to find out the basic of this form that could travel at the speed of light.

Mars, Jupiter. The astrologers and scientists say there are similar landscapes that appear to show that there used to be rivers on mars.

Let us look at the above statement, and the position of these two planets. We can Also include our moon in theses scenarios. Our planets appeared when the second generation was born. The moon and Earth were created at the same time but they came later. We must include the sizes of these planets and their distance from the sun. The reason for this as any one can see are they will be affected differently. One thing in our favour we do not need to go into any ridicule's theories. Let us forget for a minute that there was a creator. Let us take a peek at Mars. It would have the same structure as the earth. This means that it would Also have water and an atmosphere. Being So close to the sun the water would Soon evaporate. This means in no time there would be no atmosphere. This should mean that the crust would form quicker and it would be a lot deeper. The moon being smaller than the Earth and Mars should in fact Also have cooled down quicker. This in fact appears to be true.

We are now going into the realm of a creator on the next pages. Nothing to long that will make you lose interest.

I BIT ABOUT THE BIBLE

Why can all happenings that the bible have reported are now being explained by geological findings. There is no way that these findings can be invalidated.

And God said let there be light. After a while nothing had happened. Let there be light. God waited. Still nothing happened. After another attempt, and still no success he was starting to get annoyed. This is the last time I am going to try.

I don't know why you are getting annoyed EdiSon said I have had 4,999 attempts, I will try once more then I am giving up said EdiSon. God as everybody should know is not of human form. This is shown throughout history. He destroyed everything on Earth numerous times. This accrued when he flooded the Earth more than once. Then there was the Egyptian army. He Also destroyed Sodom and Gomorrah and three other cities Also the first born of the Egyptians. That to me means he was not of human form. You can understand why if he had showed himself in human form it would not have made him likable to all humans. As we all judge people by how they look. Were and how he became to be is beyond the human imagination. This will open your eyes to why there is a god. It explains in an easy to understand language that anyone can understand. There is no other way that the Earth and all the matter in the universe could have been created by Some form that came from outer space. Or from microbes, or from Earth's sea creatures, and then there is Also apes, bacteria etcetera. There is one thing for sure is that it was before there was anything in space. The only other explanation is God

Einstein said that he wanted to find out the basic of this form that could travel at the speed of light.

God as everybody should know is not of human form. This is shown throughout history. He destroyed everything on Earth numerous times. This accrued when he flooded the Earth more than once. Then there was the Egyptian army. He Also destroyed Sodom and Gomorrah and three other cities Also the first born of the Egyptians. That to me means he was not of human form. You can understand why. If he showed himself in human form it would not have made him likable to all humans. As we all judge people by how they look. Where and how he became to be is beyond the human imagination. This will open your eyes to why there is a god. It explains in an easy to understand language that any one can understand. There is no other way that the Earth and all the matter in the universe could have been created by Some form that came from outer space as microbes, or from Earth's sea creatures, apes etcetera. There is one thing for sure is that it was before there was anything in space. The only other explanation is a creator.

Einstein said that he wanted to find out the basic of this form that could travel at the speed of light.

There is a theory that is being kicked about that Some stars are created by gases getting hotter and coming together to create a star. If this is the case then these gases must be positive to allow the gases to come together. Does this mean that in time there will be more stars in the universe and when they finally cool down to the state were they will finally explode, as in the Big Bang, and lots of little bangs? This would mean that the matter in the universe is increasing. Sorry again, the matter in the universe is not increasing. It changes its composition between gases Solids and plasma. Please note when I say Solids this includes dust. If there were elements in the universe that could course an increase in the volume or mass, then over the last fifteen billion years then just think how the universe would look now. We would must go back to the Big Bang. The sun would must be the start of this statement. Bang.

As the universe expands then the appearance of gases would appear to vanish. Does this mean that they are being drawn together or that like all other matter they are moving apart? This is a fact as proven by astrologists. It is expanding. But I will shake you all up in time there is no universe. What have you done with it I can hear you scream? Do not panic you are all right now?

A laugh a minute. The problem is it's the minute after every three hours.

They say that there are planets that have the same landscape as Earth. They presume that because of this then these planets must have had water. And therefore there are mountain ranges and valleys. I have tried to think about this and use the small amount of knowledge I have about space and the Big Bang. I do not know if I am right in my way of thinking but here goes. As we know by the Hubble space ship and space ships going out to investigate the planets landscapes that different planets have different landscapes. We know there are gas planets, ice planets, and rock planets. I disagree with their observation of gas planets and I will say why later. The Big Bang hurled the Sun's contents into outer space. These would consist of fire balls, gases, and rock formation. The fire balls would I imagine stay in a near enough to a sphere. The Solid ones would Also be as a sphere, but basically their surface would be uneven. These would appear in a crater form with a splatter of mountains. As I have learned when in doubt go back to basics. The basics are the Earth's construction. Basically, the Earth's construction is the same as other planets but each one is at a different time in its evolution. Let's look at the plates on Earth and how they formed, by the bible's definition. God brought together the land and the water. Time passed (billions of years.) before he intervened again giving life to the planet. Before this, what stages would the Earth's land mass have change. The centre of the Earth would be of molten plasma. There would be a land mass of twenty-five%. And the remainder would be seventy-five%. There would must be places between these two masses that released the pressure. We know for a fact that volcano's and Earth quakes are where the pressures are released. The volcanoes as we know in time Some become dormant. And in the end, they are covered by vegetation. The Earth quakes leave a range of mountains. The rains fill the valleys that have been created by these Earth quakes. In this scenario the mountains haven't be carved out from the rivers. The information found out by archaeologists bedded onto the rocks would be the same. That to me concludes that the planets that appear to have had rainfall to carve out these mountains no longer apply. If this is the case, then there would be no glaziers All of the planets in the universe that were the result of the Big Bang none of them contained ice. This came about as they travelled further and further away from the original Sun, and every other bang. We do know that the universe is expanding. That to me means that they are getting colder. And as each sun explodes all matter is getting smaller. That Also means that the sun's are Also getting smaller. This in time will mean that Solar system will Also get smaller. There will come a time when all the sun's and planets will no longer be seen. In fact, if you take this as a fact then everything will finish up as a gas and dust. When I say time, this is beyond the imagination of us mere humans. There is another scenario that must be taken into consideration. As the sun's explode after every explosion the pieces that are hurtled away get smaller and smaller this covers sun's, asteroids, and planets whether this would affect the force fields So that they would not be as strong. I should imagine it would. If the sun's are not as large then they would not give out the same amount of heat. As we know that the universe is expanding we can Also deduct that one day it will disappear. This has given me another thought. The scientists say that there is no object or force can ever in time that it no longer exists or can ever reach its largest size. The largest thing in this universe was the original sun. Who is going to disregard that Gods creation could ever be superseded.

Mars has about one third the gravity of Earth. A rock dropped on Mars would fall more slowly than a rock falls on Earth. (I wish we had that gravity here on Earth I would have saved thousands of pounds catching things that I have dropped and broken). A perSon who weighs 100 pounds on Earth would only weigh about 37 pounds on Mars because of the reduced gravity. (This would Also be a woman's idea of heaven to lose So much weight in no time.) Mars' atmosphere is much thinner than Earth's. The atmosphere of Mars contains more than 95 percent carbon dioxide and much less than 1 percent oxygen. NASA's Spirit and Opportunity rovers found evidence that water once flowed on Mars when they found minerals that only form in water. NASA's Mars Global Surveyor orbited the Red Planet for nine years. The Orbiter found that Mars once had a magnetic field like Earth's that shielded it from deadly cosmic rays. The reason for this is that all planets would have been of molten plasma as they were from the sun's plasma. And as they travelled through space they cooled down, the reason that Mars has a thicker crust is that it is further away from the sun than the Earth. That means that it was more susceptible to the absolute zero temperature. The largest of these plasma balls would end up being the sun for all the other planets, and So create our galaxy. The reason why NASA'S Spirit and Opportunity Rover say there is proof that at one time Mars contained water. The position of Mars at the last observation shows to me that it is now, and as time goes by it will get colder. By working on logics, and Also the findings of what the scientists say that the Big Bang did accrue. When this happened, Mars would have been blasted out into space. So far So good. They say they have seen that on the surface there were signs of volcanic eruptions. This then means that mars has a molten plasma contents and Some rock contents. In time the same process would happen to earth as it travelled further and further into space the surface would cool down. What is going to happen to the surface of Mars? The circumference would start to cool to the extent that a crust would form. The next process in this scenario is that in time most of surface would be covered by this crust. There would be pockets on the surface that would have release valves as in volcanoes. As the process continued the coldness of space which is absolutely zero would still be cooling the planet down but now Some thing else would Also happen. The heat that was still available on the planet would cause the cold air to turn into water. Now we have water cascading down off the mountains and this would form rivers. Perhaps the most significant discovery regarding the Martian surface was the presence of channels. What is So meaningful about these channels is that they appear to have been created by running water, and thus providing evidence to support the theory that Mars could have been much more similar to the Earth at one time. All the planets in our galaxy would go through this process. A surface feature that has remained in popular culture since its image surfaced is the "Face on Mars." When this photograph was captured by the Viking I spacecraft in 1976, many took it to be proof that alien life existed on Mars. However, subsequent images showed that lighting (and a little imagination) are what brought life to the formation.

The sun's light rays are white because it is the colours that make up the white sun-light.

RUBBISH

Sorry this does not compute.

The prism experiment shows that the colours are part of the sun-light. This is only a reflection of the colours. The colours are made by the gases. Each gas/colour is produced this way. These can be seen around our sun. As you will notice it is around the circumference, they can be seen in the distance on all occasions. And there is a time when you see them at the bottom of waterfalls. You will notice that when you see them at the bottom of a waterfall there is Something I have noticed. The waterfall cascading down Also causes a spray that travels in an upright direction that covers the existing rainbow. This does not change the colours to the rainbow this seems very strange to me. The scientists say the reason we can see them is because the light hits the raindrops at a certain angle, this is then transferred to our eyes. This can Also be seen here on Earth you get these sighting lots of times the reason you can see them is because you are looking at them at a more condensed angle. What is the reason for the colours being in the same order? The answer to this is because each gas is of a different weight. You may ask then why are the colours shown in such a way that they are So uniformed in their appearance, as in a ring effect. All gases rise as we know. The spinning of the planet or the sun creates a force that holds all things belonging to that planet or sun in its force fields. I dispute the theory of gravity drawing everything in. if this was the case then everything including gases would be at ground level. So we know that the gases are of different weight, So they would have a different position in their orbit around the planet that was their home. Now these experts usually have theories, which they would say take with a pinch of salt.

I am going to agree with the scientists in what they say. You might ask what it is that I am agreeing to. Is it to their findings that the sunlight hits the raindrops at a certain angle, and that is what produces the colours of the rainbow? That must mean that the raindrops contain oils or more likely gases, and each gas produces a different colour. If this is So then they must Also contain the abilities of a prism. That is what the experiment of a prism does. Now comes my big question. There are rainbow colours around our planet and sun's. The red colour is the strongest, So I take it this is the lightest. That is why it is the outside colour. And here is another conundrum, why do all these colours follow the contours of the sun's and planets. But on earth it is in a bow. If there was no ozone layer around the Earth, then there would appear the same sequence of colours. We know that there are all these gases are above us because we are burning the components that make them. The part that we see inside the sun's centre is shown as a white light. All around the sun's circumference there are the colours of the rainbow. One question, why do we see a white light, when they should be colours.

(ProfesSor Cox in the TV program about space, lifts up a small stone, says, gravity is one of the weakest elements in the universe, could he please come and move the tank in my yard.) The reason he can lift a small stone So easily if because of Isaac Newton's deductions on how gravity works. The reason he can lift a small stone is his strength/force is greater that the weight of the stone.

There is another theory that the gravity in a black hole is that strong that nothing can escape from its gravitation pull it is So strong that it is pulling in all the surrounding matter, and here goes it is So strong that not even light can escape from it. Come on who's pulling my chain. What force can possibly keep light in. this must be one of the worst theories that has ever been put forward. (Answers to follow later.) That is if we have not been dragged into a black hole.

MOSES PLAGUES

The first plague was the water being changed red. This is explained by scientists as are all the miracles can be explained by forces of nature. It was not just red it was changed into blood that is why it was red.

Gases in space ignite and this produces another planet/star.
Rubbish
Let's blow them away. answer the get blow out into the outer edges of that galaxy Solar system gases that are near the sun's explosion get caught in there gravitation, as in Saturn.

There are three reasons which prove that there is a God. If you don't know the reason you will find the answer's in this book.

Einstein's pi.
Child's play.

Andromeda and the Milky Way colliding no chance!

Adam and Eve were taking a walk around the garden.
'There does not seem to be much to do around here Adam have you any suggestions.
'Let's sit down on this rock and ponder over that. I cannot think of anything to do; the garden takes care of itself. Yesterday we went fishing but after a while you get fed up catching fish every time your caste in. and I was getting fed up with fish for dinner tea and breakfast. The vegetable garden Also takes care of itself.' we have all the fruit we can eat; I suppose we could have Some hanky panky.'
'I know Something we haven't had Adam, as they passed an apple tree in the centre of the garden. What about eating one of these apples?

'We cannot eat the fruit from that tree,' Adam said, 'God has forbidden it. He said that it is the tree of knowledge.'

'Surly it can't be wrong to have Some knowledge, how are we supposed to learn anything. Come on Adam try an apple God is not here So he will not know. Go on don't be a spoil sport. Eve pulled an apple off the tree and handed it to Adam.

'He took a bite and handed it to Eve who took a bite Adam began choking on the apple and he could not dislodge it. This is one of the reasons given by the bible for man having an Adams apple.

'It's not that bad,' Eve said. Adam what are those funny things that are hanging there?' I don't know what they are called.' Eve we are naked quick let's hide.' they went behind Some bushes and covered their nakedness. It was not long when God appeared. Where are you Adam?' God said.

'We are behind the bushes,' Adam said.

'Why are you hiding?' God wanted to know.

'Because we are naked,' replied Adam.

God was annoyed with Adam. 'You have eaten the fruit from the tree of knowledge, why when I told you it was a forbidden fruit,'

'Eve said that I should eat the fruit as you would not know.'

'Don't you know that I can count, I was bound to notice that one was missing in time? Because you have diSobeyed me you will banned from the Garden of Eden.'

'Does that mean you are sending us to hell?' Adam asked.

'No you are being sent to a place that one day will be called Milton Keyes. And your crops will be plagued by parasites, and weeds will be abundant amongst your crops. Eve will bear children in pain as a punishment.

Now let us look at what has happened. We are still being punished after all this time, and women suffer when giving birth. There is another scenario to the banning of Adam and Eve from the Garden of Eden. Have you thought what it would mean if Eve had not enticed Adam to eat the apple or the apple tree was a crab apple tree or a baking apple? Adam would have spit it out and we males would not have an Adams Apple. But no what accrued was Eve made Adam eat the fruit. This means that if this scenario had not happened then there would have been no one else on Earth. That includes you and me.

This must mean that if Eve was going to bear children then this must be the time women started to have periods. For those who do not believe in a God, but that we evolved from microbes from space or from the sea. And then we evolved. If this was the case how on Earth did we manage to multiply?' this brings another scenario. Woman would be the first thing that meat predators would seek out, as they can smell blood from miles around. Our species wouldn't have lasted five minutes on this planet. Animals do not give birth in pain, or have periods, if they did how long you think they would have survived. The evolutionists compare us to the ape species as our nearest relatives *I* am Sorry I have looked at the ape/monkey families and I can see a slight resemblance. Okay So once I picked my nose and scratched my bum. Okay let's look at the scenario that we were related. Can you think of a time that when we decided to evolve and usually evolving means to improve? So the female species suddenly started to have periods, and not only that but let us scream out in pain when we give birth that will surly frighten the predators away. What we could do is hide behind a big bush shake it like mad and then scream. Every species that evolves usually do So to survive. There certainly is a big step from ape to human. We can dig up dinosaurs' bones and to get nearer to mans time on this Earth, how is it that with the billions of people that have died each year there does not seem to be any evidence. Can it possibly be that the old bible saying is right? From dust to dust. As I have just said that when a species evolves it usually means that it benefits from it. There is one thing where man has not benefited from evolving in the chain of events and that is their arms lengths. I can hear you saying what the hell are he talking about. Has there ever come a time when you have wished that your arms were longer. What about when you need Something of a high shelf. Or if you are like me and you have dropped Something and bend down to pick it up. Sometimes it takes that many attempts you say to yourself drat it I will leave it until tomorrow. There is a more beneficial time when this scenario applies. Fastening your laces. Still I have overcome that scenario I now buy shoes that have a Velcro fastener. And how many times have you bent down to pick Something up but it was further away than you thought. You try again no missed it.There was a point that I wanted to bring to the attention of all the bodies interested in the bible. Maybe I should have said to the people that do not believe in a creator. I didn't know where to put it in the above scenario, So I will ask it now. It concerns man having an Adams apple. I have not heard one explanation from all the boffins that are interested in one way or another. Why does man have an Adams apple? This is really going to get in their craw. What explanation can biologists give? Come on you experts spit it out.

Darwin came up with evolution

This was by studying creepy crawlies, animals and other forms of wild life. He realised that there was certain traits and similarities between living things. He said that they had evolved over time. Let's have a look at what he meant by time. When he says time, he meant over billions of years. This means in the beginning we must have come from bacteria that lived in the sea. They could not evolve from anything more intelligent because what form of life could they have come from. So these species of bacteria lived in their own environment. I take it, it must be the sea, or did they start out on the land and then evolve into sea bacteria or was there two types of bacteria and each species evolved in their own environment. Everywhere on the planet it has its own bacteria whether it is animal vegetable or mineral. I have introduced mineral, but bacteria are on the surface, but it does not eat the mineral if it did over the fifteen billion years there would be no minerals left. tongue in the cheek for this one. This is what the experts say. It is the bacteria that eat away at all living things that have died. Let's look at the easiest one that we can understand us. This Also applies to vegetation. If bacteria was the first sign of life on the planet I would like to know what on Earth could they evolve into, and how long would it take them to evolve into Something that was more intelligent. Let us presume that there were different types of bacteria, as we know there is. How long was it before they decided to evolve billions of years, or right away? But when was right away. There had to be a beginning. I mean what has the bacteria evolved from. You can't get Something from nothing. And if this is So then what did that evolve from. I could go on but you get the idea. Now we have minerals. Let us take the biological expert's point of view. The bacteria that lived in the sea. I *take* it that their supply of food must have been there before the bacteria. I will enlighten you as the food supply gets smaller than the bacteria can change chemicals into food.

Now let's really upset the apple cart. Where did the plants evolve from? And to go back even further where did the food come for the bacteria to live on. They must have been here first (Some sects believe we came from outer space, the only thing that's in outer space is what's between their ears. I bet they think the world is flat.) Let's think about this, everything must have Something to eat no matter what it is. So did these intelligent plants come from outer space on their own, maybe they were runner beans. Have you ever thought about weeds, bugs, bacteria and insects? Let's do away with the idea that we came from outer space. It must have been a big space ship. Not only would it must transport us but Also plants, trees and all the sea creatures in fact everything that is here on Earth.

We now do have one thing the experts do not know. Where do we come from? Let's look at the point of view from different people whether they are from different professional bodies or from people that have picked up the ideas from sects that want their money. You can get people to believe anything that are farfetched if you catch them at a vulnerable time in their life. We will first look at the sect that believes that we have come from outer space. I think that most of the space is between their ears. Why do I think this way you might say that I have a lot of space between my ears? I will explain this later as this answer applies to other aspects of where have come from. Charles Darwin theory is basically tied to the previous question that it all started from bacteria in the sea. And that was the start of evolution. Okay we have the plants and the bacteria let's take it from there. Was there only one or a male and a female? Or was there one two or thousands of this bacterium. Maybe there appeared billions that happened all at the same time. To reproduce there must be a form of male and female. We now there is a fantastic way that different specie reproduces. But don't forget this bacteria has not had time to evolve So we must take it that it is at its first stage. Miracles must have happened for it to evolve. Now consider this bacteria starts to evolve. They are all having a conference to see which ones are going to evolve they all can't evolve otherwise there would be no bacteria. We have bacteria So to me that means that they have not evolved. Just think about this what possibly could stop them all from evolving. Usually when a species evolve is to improve its life standards or to survive a change in its environment. I'm stuck. If the scientists and naturalists can't figure it out, then we are stuck with two questions. This would mean that once they started to evolve they all would have evolved and there would be no bacteria. And as that species evolved there former selves would die. That means that there would be only one species on the planet. They talk about the missing link between apes and man. Evolution usually takes millions of years. There is not one missing link between man and apes but millions. There is one very important thing in all this evolvement that not one of the biologists has thought about. It makes all the theories that the physicists and scientists have put foreword to be wrong. Most of my answers to all these theories are first based on logic, the main reason for my observation that man are a species of their own is based on what all the biologists have been saying. The crunch of this observation, which proves all the theories put forward is a load of twiddle twaddle, the one thing no one has thought about, and it goes on the fact that all scientists and biologists knows for a fact. I cannot understand why none of them has thought about this. The only reason that I can think of is that it will put all of them out of a job. Maybe after they have studied the facts they will see the light. (Answers later.) It takes us too are nearest relatives, the apes and orang-utans. When I was a child I loved climbing trees So that must be my instinct that came from when I originated from the apes. I am now sixty years of age and I have no desire to go climbing trees I really have started to evolve and it's only taken me fifty six years to evolve not millions of years. I wonder in another twenty years when I go senile if I will find myself having the urge to shin up a tree. Does this mean my evolution has ended and I am returning to my ape instincts? And if I can get too the time before this then I should be on my hands and knees crawling about. I hate to think beyond this. Just a recall on what has been said. For all these things to survive they needed food, So when they left the sea it couldn't have taken millions of years to survived they must have adjusted in a short time. When the sea was their home they were protected from the cold and ice. Arriving on land they must surly have been hungry and adapted right away. How have I come to this deduction, there are dung beetles that were brought from Africa to Longleate the safari park. From the time it arrived it was fed on fruit. It survived on fruit and after three years when they tried to feed it dung it refused to eat it. Maybe this was because it had no female to impress. This insect didn't take thousands of years to change which goes to prove there must be a lot more insects and animals that changed their diet habits when there environment changes, And not over millions of years. Let's go back to the dung beetle. Let's say the safari park introduced a female dung beetle. Can you see this male dung beetle following the elephants with a bucket and spade, I don't think So? Would there be any need for it to change its ways. I don't think So. Let's look at what would happen as Soon as the male dung beetle saw the female dung beetle. All he had was fruit. Well he thought I had better use what I have at hand. He rolled a big ball and introduced it to the female. The female approached with caution. There is a very unusual smell she thought. She moved closer and took an apprehensive sniff. That's good where have you been all my life she says. Now let's take a different scenario. The safari park introduces a few more male dung beetles. The other dung beetles have not been introduced to fruit So they all go about chasing elephants all around the safari park, with their bucket and spades. Now they introduced a female. Competition, now the sparks will fly, or maybe it's the fruit, I use the fruit as an example as the other scenario may smell a bit. All these beetles roll their offering to the female beetle. She inspects all the balls and goes back to the fruit one. There could be a market for this one she says. I could save a fortune on deodorants. The point is it did not take millions of years to evolve, when it comes to the crunch if the environment changes quickly the wild life must change, or the species dies out. Here's a question why has the giraffe such a long neck. All the grass eating animals get their food from ground vegetation. Where is the link between these animals and the giraffe? If there was a lack of grass then there would be no grass eating animals left or did they gradually start eating food from a higher plain. I am no expert, So I will leave it up to them.

THE ROYAL ASTRONOMICAL SOCIETY

They say mass exterminations over the last 260 million years have been caused by comet and asteroids showers. There is Also the theory that large volcanoes erupted, and this caused a massive cloud that encased the whole of the Earth. This prevented any sunshine to reach the ground. That caused everything to die. If asteroids and comets destroyed the dinosaurs then it must have destroyed everything. That means the cycle of all that was on the Earth before this must start from the beginning again. Now let's look at Some facts that the "experts" say has happened. They point out that there was an ice age. That puts a shiver down my spine. If this was the case, then this must have destroyed all of the vegetation. Which in turn must have put a lot of insects, and animals on a diet, this diet wouldn't have lasted long as they would have died. Now I am no genius but that must have affected the bird population. The experts say that there must have been no survivors. This means that the world as we know it must start all over again. How all the Earth's vegetation and creatures keep rising from nothing amazes me. If all of these species managed to live through the ice age then surly there must Some survives in the ice age and if they were going to evolve well they had sixty million years to evolve every time they were exterminated. So where are they? I can't see any at the artic or the Antarctic. I would surely have thought that it was more than likely that none of this happened and it just evolving from the first time life was introduced to the Earth. My view is that in the beginning i.e. after the Big Bang all the matter was at that time a lot larger than it is now. This is because the following bangs decreased the size of all the matter. Another point is the planets had no protection from these missiles. If this was not happening then all the planets would be hit by planets larger than their selves. At this time 260 million years ago there was no life on any planet. So it could not be destroyed. Who's a clever boy then? They say it's sixty million years ago when the dinosaurs became extinct. Let's do Some calculations, I am no mathematician but here goes. 260- million years ago there was nothing, and the dinosaurs appeared. Then they were made extinct. It took another sixty million years for them to be back and So it goes on. By working this out I forecast that the dinosaurs are ready to make another appearance. (Take this with tongue in the cheek.)

Let us see what would happen if a gigantic volcano exploded. Molten plasma would spew out all the surrounding country would be devastated. Sorry this would not happen. Plasma would not spill out at an equal force. It would overflow at the weakest part of the volcano.

We have all seen films of a volcano erupting So you must know this as I do that what I have said is true. Now let's look at what another scenario produces. Its smoke, Soot, and flying ashes Also boulders of different sizes. The heavier matter drops not at such a great distance from the cause of the eruption. So the final catastrophe of this eruption, is that it is going to destroy the Earth's inhabitants, and all living things, whether its vegetation, creepy crawlies. Some people would love that no creepy crawlies. Now let's look at how this will affect the Earth and our life as we know it. As with the plasma flow then the same will happen with the smoke, dust, and the plasma. If there was no wind and the heat from the volcano took the smoke straight up into the atmosphere or above into the stratosphere. Let's look at the first scenario; the wind would take the dust and gases in the direction that they were blowing. Then in time the wind would disperse the said smoke. In time the rain would carry the dust and other particles down to Earth. Okay So the Earth would be covered in this dust So let's say nothing can grow for a while. Just let's pause for a minute and go back. I would presume that most of the damage would have occurred around the local area. And what I have seen or read about it seems that the following spring things start to grow again. In no time the insects and small animals would be the next to arrive. There was another scenario that was much more devastating, and all the Animals are back. They have become used to their environment and are able to live with the hazard, it appears that they can live with it and even have no effects from the great catastrophe that occurred. There is plant life, insects and animals living with radiation that occurred at Chernobyl. This has not taken billions of years I mean to say radiation that must be one of the worst scenarios. To get back to the volcanoes dust cloud. It has been dispersed and the further it travels the less dense it becomes. When the scientists and weather experts say in the forecasts they make about the world being destroyed is most unlikely owing to how are weather system works. The same scenario would happen if it was an asteroid. But if it was that big there would be no Earth or asteroid left. Why is there no chance of a large size asteroids hitting the Earth and destroying it? There are two reasons why I have come to this decision. Find out later these decisions are not based on a theory, but on facts that the astrologists, physicists, and scientists have been saying for yonks. (Means a long time) I will reveal these facts later. do you think that the Earth will be destroyed by one of these things, we know by the satellites photos that the wind that move around our plant might change but whichever way they travel it usually in the same direction until the seasons change and then So do the winds? As can be seen there are lots of places that are not affected by the winds in the same way as other parts of the Earth. That to my calculations means that all the Earth's living things will not be destroyed. To go back to asteroid hitting the Earth and the sun's light being blocked out. This must have been the collision that produced our moon otherwise we would have had two moons. (That would have made a mess of a lot of love Songs) let us think how our Earth is constructed. Its round that was easy for a start. Then we have the wind that was easy. Duck there's an asteroid coming. The dust rises. Let us say it covers half of our planet. We now must look at a volcanic eruption as this has the same affect. On the first eruption or an asteroid hitting the planet the dust rises as stated. The volcanic dust then starts to disperse. It is blown by the winds. Here in England to my perSonal experiences we have had sand storms that have deposited sand here in England on two occasions and these have come from Egypt. There are sandstorms nearly every day but we do not get the sand here. This would be the same scenario as a dust cloud. There might be Someone that will come up with the scenario that sand is heavier. The dust cloud from this massive impact would not be just dust but there would be large chunks as in boulders. These of course would drop to Earth quicker. This would mean that what was left after all the larger matter had been deposited on the ground all of the dust in time would of course settle down on the Earth's surface. Also the rain would carry the remaining down to Earth. Once an asteroid had smashed into the Earth's surface, then that would be that. But a volcano keeps sending its plasma, rocks and dust into our atmosphere. We have had a volcanic eruption not So long age in Iceland. Wikipedia information. On March the 22nd a sudden rise in the water level and in the temperature by 6 degrees ((11 'f) over a period of two hours. shortly after the waters and temperature rise it fell. 18 03 2010 April 12th, 2010. Six days to May. Over October 20 countries airspace closed most of Europe airspace. The ash was high enough for it to enter the jet stream.

SPINNING PLANETS

Spinning planets and spinning sun's, why do the asteroids not spin and this Also applies to comets. They are a lot smaller than planets, So they must have been harder to hit by other matter, but it would not take much to send them into a spin. The astrologers have said that they kept crashing into each other if this was the case then how are they different from Something bigger. On the other hand, it should be a lot easier to make them spin. They should be spinning faster. Planets, sun's, asteroids and comets do not crash in to each other if they did over the time since the Big Bang the universe would be like a sandy beach in the Sahara Desert.

Our neighbouring planet Venus is an oddball in many ways. For starters it spins in the opposite direction from most other planets, including Earth, So that on Venus the sun rises in the west. Not that it happens often: a day there lasts a little more than 243 Earth-days, making it longer than a Venusians year, which is only about 224 Earth-days long. Many scientists believe that the long days are a result of the sun's strong pull on the planet. (Mercury, which is even closer to Sol, has fairly long days as well: three for every two Mercury-years). But scientists are still puzzled by Venus's retrograde, or backward, rotation. Now a team of scientists from the French research institute Astronomies Systems Dynamiques have proposed a new explanation, published in this week's issue of *Nature*.

Current theory holds that Venus initially spun in the same direction as most other planets and, in a way, still does: it simply flipped its axis 180 degrees at Some point. In other words, it spins in the same direction it always has, just upside down, So that looking at it from other planets makes the spin seem backward. I am lost with this explanation. I am going to must think about this. Scientists have argued that the sun's gravitational pull on the planet's very dense atmosphere could have caused strong atmospheric tides. Such tides, combined with friction between Venus's mantle and core, could have caused the flip in the first place. (This is not my way of looking at things. it will be explained later.)

Now Alexandre Correira and Jacques Laskar suggest that Venus may not have flipped at all. They propose instead that its rotation slowed to a standstill and then reversed direction. Considering the factors mentioned above, as well as tidal effects from other planets, the team concluded that Venus's axis could have shifted to a variety of positions throughout the planet's evolution. Regardless of whether it flipped or not, it is bound to settle into one of four stable rotation states two in either direction. The researchers add that Venus would be more stable in one of the two retrograde rotational states. So in essence, it was just a question of time before Venus started spinning the wrong way. There has been another theory that the scientists, and physicists have thought up. That in time the universe will start on receding path going back to its original beginning. This is being called the crush effect, this is the exact opposite to what other scientist are saying. I think I will put Some logic on the subject. Now only one of them can be right. You must Also have come to this conclusion. We will look at the facts. We will start with the theory of it receding. What do they base this on if you put this theory on the evolution of man and beasts? And the rest of living things then we are going to stop evolving and go back into our past as apes and So on until we finish up as bacteria. And as Some people believe, (bless their Souls) we will go back into space were we originally came from. Taking a look at this crunch scenario all the matter in space must reverse the gravity of all the sun's, moons and planets. That means that there must be Something to start this tremendous occurrence. If my logic's are scientifically proven wrong then they will come up with a different theory for it. Why do I not think that the universe will reverse back to the original? What is my logic about the planets and sun's spinning? Let's start at the beginning. I am going back to the Big Bang. When this accrued all the matter was sent out into space. There was no gravitational force to put them in an orbit, So they would continue in an outward direction. None of these objects would spin. If the physicist or scientist could explain this I would be grateful. My explanation is this miracle can only be explained by there being a God. As time is passing by and more revelations are being discovered about the bible then it gets more and more likely that there is a God. This is no more a miracle than the billions of miracles that have made this Earth the way it is.

REPLY

I hope that Some of these experts will take notice as I keep saying let's get our feet back on the ground by this I mean our planet Earth.

What about the original sun'
It is not possible to just appear out of the blue, no pun intended. It cannot be a pun because there was no blue in space. So how can the scientist or physicists explain to me where it came from I think I heard an explanation that it was created by gases in space. Here comes my question. Where the hell did the gases come from? Why do they keep coming up with different scenarios? I will try again where did the hell did they come from? Let's try again. God.

f the speed of light is the fastest thing that anything can travel at then all that was said is wrong. This is logic. This is one of the things that I dispute about size and weight. So does not light speed prove it? There is more later about the speed of light. There are two things that I have noted. So to start we will go back to the very beginning. I suppose you could say that this proves me wrong as space has no end So there must be no beginnings.

The cost of all the exploration of going into space is tremendous Benefits? Practically nil.

If we had never gone into space is there anything that we would have missed I wonder what the scientist/physicists say. Has there been anything that has been discovered in space that has been of any value to anyone on our planet. If we look at another scenario is there anything in space that would have been disastrous to this planet, if we had not gone into space. What about the trillions of pounds spent on space expiration? If we were to ask the people today would they agree to spending all this money on going into space could they find a better use for it okay, when I talk about space I am not referring to the space satellites that are being used for our benefits these are still governed by the Earth's Penny gravitational. I am talking about any planet. What has been gained by man going to the moon, nothing. It's just for prestige. I can just imagine there being a big meeting of all the dignitaries. Okay the chairman says we have walked on the moon what can we do next we must think of Something otherwise we will be out of a job. Great we landed on the moon. Now comes a big decision for the scientists. After all the flag waving and things started to settle down they started to think. Let's go and think of Something else. How about Mars? That is our nearest planet. We know we cannot land there owing to the terrain. We could say that we have found that life can live there. That's a big disappointment nobody's there. Where's the next planet. There is no point in sending more space ships around planets that we have already gone too to take more photos of loads of gases spinning about. And what is the point of landing on a planet that has nothing to offer only more photos. Has anything been found up to date that has been of any use to anybody on this planet. All this money should be used to improve life on this planet. Can be used to discover great things on this planet, say finding more cures for our diseases and global warming. Just think of what could be achieved if this money was spent on more human reSources. Man has only advanced from innovations and scientific discoveries that have nothing to do with space. What really keeps this expiration of space is they keep moving the goal posts back. I would have thought the first place they said that was likely to have aliens on was on the moon. Then it was the next nearest planet. Then it was Mars. Now they know that there is no life there they have found another planet that they say could have life there. The only reason for this is to keep there-selves in a job. As it has been said before the Lord said the heavens belong to me. Let's take a inventers point of view. Once you have invented Some-thing or made a scientific discovery you don't do that again So that means there is only one planet with life form on, the Earth. The Americans have at last come to their senses on investigating u.f.o.'s they don't bother. So why are they still spending billions of dollars in expiration of space. The world is flat, I wonder if any one believes that nowadays. Then there was the much to do about it was all an American put up job that the Americans had landed on the moon. All these outlandish stories are just So that the news media can sell papers and magazines. When the Big Bang took place, I believe it created the Milky Way. Or was the Andromeda galaxy the first one. If this is So then the original Big Bang created all the planets, sun's, gases, asteroids, and comets. That to my way of thinking is that all this matter was smaller. So how come our moon has a mass that is bigger than it should be considering our Solar system is the last to appear. How can this be it is not logical whose messing about my logical brain. I feel like my head is going to explode and to me when it explodes to me it will be greater than the Big Bang.

TIME AND SPACE

When trying to work out time and distance in space the physicists are mixing up the two different equations

As with all moons they show that any interference with their surface by a meteorite or a volcanic eruption. The scenario of these two scenarios would have a different effect on their surface. A comet or meteor would have one scenario and a volcanic eruption would have a different effect. Let us take the first scenario of a comet/asteroid there would be a crater and that crater would have at the rim of this crater would of course be a lot different than a volcanic crater. A volcanic crater would have a deeper crater and the tip would be higher. They say that our moon is the largest satellite in our Solar system, how can this be possible. all the scientists and physicists know that when the Big Bang accrued, (this was from the original Sun exploding.) the average perSon knows and he doesn't must be an expert on physics, that if Something explodes then from larger objects you are left with a lot of smaller units.

If there is any likelihood of another planet having life on it then surly the most likely planet would have been the moon. What are the odds of a sun exploding and out of all the material that was flying around the moon had no plasma in it or So little it would not make any difference to its behaviour. Let's look at this in a new light. No pun intended. If the moon was constructed in the same way as our Earth then the gravity would have held any gases in its gravitational pull. There are no gases So either there was none, or they escaped. The second scenario is most unlikely. The moon is the closest object to Earth and Earth was constantly being bombarded by meteorites. They really reach the ground as they burn up in the atmosphere, as time goes by the meteorites will get smaller and smaller. So you don't need to worry about the Earth coming to an end that way.

They say they can see puffs of dust rising on the moon's surface. They have come up with the theory that they are caused by meteorites hitting the moon's surface. Why can't they be picked up by all our sophisticated equipment? We know they do not burn up as the moon is not protected like our planet. I do not know if I am seeing it right, but all these puffs of dust seem to appear near the top of a crater. These puffs of dust that only appear at the top and around the edges are because of the winds. The lighter dust would be blown and trapped at the top of these craters. All other moons and planets were created by the sun's exploding. If our moon is a solid it is the only satellite that has this structure, they say our moon has a Solid core. Also, all other moons, and planets have a Solid core. Asteroids are different and that is they are the smaller fragments from a sun exploding. The reason that there are small puffs of dust it there was only a small amount of lave when the moon came to be. As the years passed by the build-up of pressure became less and less. We have now reached the time when this pressure is partially nil. The reason that they appear the top of these craters is because the crust of the moon at this point is thinner. In time the Earth will finish at this state.

Maybe just maybe it came about by a divine hand. The physicists, astrologists and the scientists all agree about the Big Bang. So, the first Sun that exploded brought into being all the planets, sun's dust and gases.

God took the fire, water, and material from the universe and he created the Earth. There was no point in creating two, Earth's So he created the moon to reflect the sun's rays to light up the night sky on Earth. That is why the moon is a Solid. This can be seen by the conditions of the moon's surface. There have been no volcanoes or earthquakes on the moon. If you notice all the craters are caused by objects hitting the moon's surface. If they were caused by volcanoes, they would have a different appearance. As can be seen on our planet Earth, a volcano's eruption spews out hot molten plasma. This has the effect of creating a mountain. This is because the plasma graduals build up around the brim of the mountain. The two scenarios are entirely different in their makeup. A comet hitting a planet would leave a crater with a much lower rim, and be a lot smoother, as on the moon. When the moon was created, it contained hot plasma and Solid rocks. The plasma was only of a small amount and that is why there is a Solid inside. With their only being a small amount of plasma this was not hot enough to cause volcanic eruptions. Gradually as the plasma burned up the oxygen in its centre a Solid core was formed. Looking at the images of a volcanic impact on the moon, they look very much like a picture of a comet. As a comet is comprised of mostly iron, nickel and stone, if this hit the moon it would give the impression that it was a volcanic eruption. Owing to the moons gravity being less than the Earth's and there being no atmosphere So there would be no burning up of the comets. The gravity would not slow the comet down causing it to have less impact. The offside of the comet this being the part that does not in fact actually hit the moon, there would be less of a impact So it would not change it in its contour or in the way it looked. As the Earth was created with the following being brought together with burning plasma crust and water and the moon was mostly of crust from the heavens. Also you must take into the equation the angle that this object is hitting the planet. At a very low angle it would bounce off. At too high an angle it would embed itself into the earth and leave a crater. I wonder at what angle it could possibly be at to remove enough matter to this means that the object hitting the Earth must have more mass. This would then destroy the Earth. There is no matter that is flying around the universe that can knock a big chunk out of the Earth. So where has this boulder come from that has collided with Earth to give us our moon? This collision must go against all the laws of physics. Why you may ask? If we take a look at the Earth's contour there is nowhere that it shows signs of an impact. So here is my problem the size of this object hitting the Earth.

The Earth's spin cannot be coursed by an object hitting it. This Also means that Isaac Newton's law about equal forces comes into the equation. If you have two objects where one is larger than the other one that have a similar construction travelling at a tremendous speed what do think will happen. Let us look at the different scenarios. One thing we do know is that the both will be damaged. One is more than likely, being damaged more than the other. The smaller object is the most likely one. Now these two objects must connect at a precise angle. The steeper the angle there will be less damaged to the larger one. If they met at a ninety degree angle the smaller object would displace a small amount of the larger one's surface. It would Also be embedded in the Earth's surface. This would mean that the smaller object would have the majority of its matter/rocks showing above the other one's surface. This cannot be seen on the Earth's surface. Let us go to the extreme which is it glanced of it at a low angle. This would disturb the larger ones surface and there would be smaller objects created. And the offending object would again be buried in the Earth's surface. This means that there must be an angle, between these two scenarios were if it was possible a chunk the size of the moon would be displaced. I cannot see how any object no matter ever size could possibly achieve this scenario. 'Physitrists. Got ya.'

Bible

Moses went to the pharaoh and asked him to free the Israelites on Gods bidding. Pharaoh refused Moses, said that if he did not then the Egyptians would feel the wraith of the lord. To show Gods powers Moses threw down his staff and it turned into a snake. Pharaoh's adviser threw down a staff and it turned into two snakes. Moses snake struck and ate the two snakes of pharaohs. That's one diet that's gone for a burton. (Moses left with a larger staff than he came in with), Moses warned the pharaoh that God's wrath will be a lot more severe.

THE PLAGUES OF EGYPT

I will look at what they must say and put my two-pennyworth in. they have come up with the reason for each plague and why the last plague caused the next plague. The start of these plagues was of course a volcanic eruption. This is the reason given by geologists, biologists and physics'.

The Plagues of Egypt were ten calamities that, according to the biblical Book of Exodus, Yahweh inflicted upon Egypt to persuade the Pharaoh to release the ill-treated Israelites from slavery. My thoughts. I can understand that the regime was very severe in the punishment of any Israelite stepping out of line. As we know the Israelites had their own part of Egypt. And their numbers were considerably more than the Egyptians. They would not must queue up for their food, So they must have grown their own. Also they had to be reasonable fit to carry out their duties. Only the trusted slaves had work in the homes of the Egyptians. These would have been from countries that did not put up any strong resistance when the Romans arrived. And we must understand at this time they were not Jews, but Israelites.

The plagues served to contrast the power of the God of Israel with the Egyptian Gods, invalidating them. Some commentators have asSociated several of the plagues with judgment on specific Gods asSociated with the Nile, all the Gods of Egypt would be judged through the tenth and final plague:

Without the biblical understanding that the beginning of Christianity started in the Garden of Eden, this is just a theory as is the following statements. I do not remember where I read this on the internet, but I do not want bore my readers with too much here say. Here's what was said. With the promise of a Savoir, given immediately after Adam and Eve sinned, the Christian of today will be literally tongue-tied and unable to resist the supposedly intellectual assaults of those who do not understand the Bible and who constantly promote the view that Christianity is nothing but a weak variation of a pagan religion.

Nothing could be further from the truth.

Christianity was the first religion. All other religions, including Judaism, are counterfeits of Christianity. There is the belief that the first Christians were Adam and Eve. And they were told that there would be a messiah that would save mankind. Why would this statement be made? God punished Adam and Eve because they now had knowledge. They already knew there was a god. As for life here-after this was beyond their capabilities. But as we know all over the world every human tribe has their own gods. In the earlier days these did not have belief in the hereafter. There is no way that the above statements about Adam and Eve being the first Christens can be true.

The plagues as they appear in the 1984 New International Version of the Book of Exodus, arek

Water into blood.

The Nile turning red

The experts say the reason the Nile turned red could have been caused be a volcano erupting. This caused ash to fall; this ash would contain lots of poiSons chemicals. So what we have now is this ash would of course contain chemicals that could have had two affects one the type of chemical

So we are in Egypt. Now my geography is not top class, but I think it's a long way to the nearest volcano to Egypt. Let's look at it from different angles. One Moses has very good eye sight

This is what the LORD says: By this you will know that I am the LORD: With the staff that is in my hand I will strike the water of the Nile, and it will be changed into blood. The fish in the Nile will die, and the river will be So contaminated that the Egyptians will not be able to drink its water.

The First Plague: Ii is said that all the water is turned red. https://en.wikipedia.org/wiki/James Tissot the fact is it was turned into blood. We will now go into what our biologists say. This was coursed by a volcanic eruption. The poisons that were contained in its ash fell on the river and this coursed all the fish and bacteria to die. This then turned the river into a stinking base for red algae to increase and that is why it turned red. This popular chemistry demonstration is often called turning water into wine or water into blood. It's really a simple example of a pH indicator. Phenolphthalein is added to water, which is then poured into a second glass containing a base. If the pH of the resulting Solution is right, you can make the water turn from clear to red to clear again, if you like. This is mostly unlikely to have these elements together all at the same time. The nearest volcano to Egypt is about three hundred miles away. At what time-scale was there between Moses asking Aaron to strike the river with his staff and the river turned into blood. Did he see the cloud of dust approaching? Maybe their Met office had this information. We need them here today. Enough said. We know man can go without water for a period of three days So that must have been the time scale of the first plague. here's another way to look at things. if the ash falling from the volcanoes' eruption killed all the fish and bacteria to die how come the frogs lived. okay got it the frogs were capable of leaving the river. The fish would be in a right flap if they tried to leave. let us go back to the beginning. the ash falls. there's a drought this gives the alga time to grow.

A dirty thunderstorm (Also volcanic lightning, thunder volcano) is a weather phenomenon that is related to the production of lightning in a volcanic plume. A famous image of the phenomenon was photographed by Carlos Gutierrez and occurred in Chile above the Chaiten Volcano. It circulated widely on the internet.

Other notable image of this phenomenon is "The Power of Nature", taken by Mexican photographer Sergio Tapiro in Colima, Mexico, which won third place (Nature category) in the 2016 World Press Photo Contest.

Other instances have been reported above Alaska's Mount Augustine volcano, Iceland's Eyjafjallajökull volcano and Mount Etna in Sicily, Italy. https://en.wikipedia.org/wiki/Dirty_thunderstorm - cite_note-8

Causes

A study in the journal Science indicated that electrical charges are generated when rock fragments, ash, and ice particles in a volcanic plume collide and produce static charges, just as ice particles collide in regular thunderstorms. Volcanic eruptions are Sometimes accompanied by flashes of lightning. However, this lightning doesn't descend from storm clouds in the sky. It is generated within the ash cloud spewing from the volcano, in a process called charge separation. As the plume started going downwind, it seemed to have a life of its own and produced Some 300 more or less normal [lightning bolts] ... The implication is that it has produced more charge than it started with. Otherwise [the plume] couldn't continue to make lightning. Volcanic eruptions Also release large amounts of water, which may help fuel these thunderstorms.

Just a thought. As can be seen in the above statement it states that when a volcano erupts it creates fierce thunderstorms and torrential rain. If this was the case and we are going by what we know about volcanic eruptions this is true. Then this torrential rain would follow the ash that had been produced. This would follow the direction of the Nile. This would be happening while the ash was falling. If this was the case then we would not have had any algae. This is because of the torrential rainfall that the volcanic ash produces. Let's see the bright boys Sort that out. I suppose they will say that the lightning and the rainfall was only at the top of the volcano. It should still reach the Nile. When these say should reach the Nile. Okay we know that the nearest volcano is three hundred miles away. The eruption of Chilean volcano Chaitin May 6, 2008 there was witness that saw lightning flashes that rolled along the ground twenty five miles away from the volcano. We have been looking at the cause of volcanoes in the light of known facts. We must not lose the fact that the bible has pointed out the occurrences of the ten plagues and that it was the Lord in fact that brought this about.

The plagues in the Quran

In the view of Islam, the plagues were almost identical. It is mentioned in the Quran, specifically in Surah Al-A'raf verse 133 *"So We sent on them: the Tuwfan (a calamity causing wholesale death, a flood or a typhoon - the locusts, the Qummal, the frogs, and the blood (as a succession of) manifest signs, yet they remained arrogant, and they were of those people who were criminals».* The Quran further relates that the plagues included a mighty blast, showers of stones and earthquakes (Ali, Notes 3462-3464 to S. XXIX.40).[32]

Scholarly interpretation

The Book of Deuteronomy, in which Moses reviews the events of the past, mentions the "diseases of Egypt" (Deuteronomy 7:15 and 28:60), but means Something that afflicted the Israelites, not the Egyptians; in fact, it never mentions the plagues of the book of Exodus. The Exodus plagues are divine judgments, a series of curses like those in Deuteronomy 28:15–68, which mention many of the same afflictions; they are even closer to the curses in the Holiness code, Leviticus 26, since like the Holiness Code they leave room for repentance. The theme that divine punishment should lead to repentance is echoed in the prophets (Amos 4:6–12, Ezekiel 20), and the form of prophetic speech, "Thus says Yahweh", and the figure of the prophet as divine messenger echoed in the late prophets Jeremiah and Ezekiel and the Deuteronomistic history. The theme of Pharaoh's obstinacy is likewise referred to in the 6th century prophets – Isaiah 6:9–13, Jeremiah 5:3, and Ezekiel 3:7–9.[35]

Historicity

While proponents of biblical archaeology argue that the plague stories are true, a large consensus of historians believe them to be allegorical or inspired by passed-down accounts of disconnected natural disasters. Some scientists claim the plagues can be attributed to a chain of natural phenomena triggered by changes in the climate and environmental disasters hundreds of miles away. Some historians Also point to the Ipuwer Papyrus to suggest a possible cataclysmic event in the history of Egypt which might parallel Some of the incidents described in the biblical account of the Plagues. However, no reasonable natural explanation can be found for several of these (i.e. the water being turned into literal blood, unpeaceable darkness for three days, and all the firstborn - including Pharaoh's Son - dying at the same time.

Archaeology

Some archaeologists believe the plagues occurred at the ancient city of Pi-Ramesses in the Nile Delta, which was the capital of Egypt during the reign of Ramesses II. There is Some archaeological material which such archaeologists, for example William F. Albright, have considered to be historical evidence of the Ten Plagues; for example, an ancient water trough found in El Arish bears hieroglyphic markings detailing a period of darkness. Albright and other Christian archaeologists have claimed that such evidence, as well as careful study of the areas ostensibly travelled by the Israelites after the Exodus, makes discounting the biblical account untenable. The Egyptian Ipuwer Papyrus describes a series of calamities befalling Egypt, including a river turned to blood, men behaving as wild ibises, and the land generally turned upside down. However, this is usually thought to describe a general and long term ecological disaster lasting for a period of decades, such as that which destroyed the Old Kingdom. The document is usually dated to the end of the Middle Kingdom, or more rarely, to its beginning, fitting the Old Kingdom destruction, but in both cases long before the usual theorized dates for the Exodus.

Natural explanation

Some historians have suggested that the plagues are passed-down accounts of several natural disasters, Some disconnected, others playing part of a chain reaction. Natural explanations have been suggested for most of the phenomena:

Plague 1 — water turned into blood; fish died

Dr. Stephen Pflugmacher, a biologist at the Leibniz Institute for Water Ecology and Inland Fisheries in Berlin believes that rising temperatures could have turned the Nile into a slow-moving, muddy watercourse—conditions favourable for the spread of toxic fresh water algae. As the organism known as Burgundy Blood algae dies, it turns the water red.

Alternatively, a bloody appearance could be due to an environmental change, such as a drought, which could have contributed to the spread of the Chiromantical bacteria which thrive in stagnant, oxygen-deprived water

FROG PLAGUE

The Second Plague: And Aaron stretched out his hand over the Waters of Egypt and the Frogs came up and covered the land of Egypt.

This is what the great LORD says: Let my people go, So that they may worship me. If you refuse to let them go, I will plague your whole country with frogs. The Nile will teem with frogs. They will come up into your palace and your bedroom and onto your bed, into the houses of your officials and on your people, and into your ovens and kneading troughs. The frogs will go up on you and your people and all your officials.

The reason for the plague of frogs was that they left the river. This was because the water was poisoned all the food supply that the frogs ate Also disappeared. This is quite a reasonable explanation as put forward by archaeologists. This mass of frogs left the waters to find they had jumped out of the pot into the fire. They covered the land of Egypt. They should have collected them and sent them to France were there was a big a demand for them there. In no time at all they started to feel peckish. After a while the frog's food supply would run out. We now have a lot of dead frogs. If we recall after three days, the water had to return for man to survive. As can be seen by the above the ash clouds would have produced storms. Down comes the rain and everything goes back to normal. Surly if this was So then any remaining frogs would be back in the swim. We must remember that the Egyptians had gone three days without water So as Soon as the water was restored back So they could drink the water they would head towards the river. When the Nile was turned into blood, all the frogs left the water and died. This then caused the second plague. Could any of the experts in all fields that they studied have forecasted this? And then they would must forecast another eight plagues. Moss's was certainly an expert in many fields. I won't go through them all as I am sure you

Bang Goes That Theory

get the drift. The scientists say that what caused this was the grain that was in storage got a fungus disease owing to the sunlight being blocked out by the darkness that followed the plague. Sorry to disagree with this formula. I can see how one plague was the cause of the next plague. We know that the frog breeding season starts in the spring as nearly all the living things are governed by the seasons. That means that the tadpole's transformation would have accrued in early spring. If the frogs were forced to leave the river then it must have been at the end of the summer months. The scientists say that the stress would have made the frogs mature earlier. If this is So, then that makes my point more reliable. What is it? Can you think what it is? There is more than thing that they have got wrong. Okay let us be patient. There had been a global warming. This meant that there was a drought. This was why the fungus was prevalent. The algae died this caused the river to flow red. (The bible says the rivers flowed with blood.) There is a big difference. How long would it take for the flies to emerge from a maggot to a fly? The weather would be hot and dead meat does not need a fly to produce maggots. The maggot can change to a fly in one day. All you must look at is the human body. If we do not keep clean what is the first thing that happens of course its head lice. Meat has a natural biological way of producing maggots otherwise where there are dead carcases that cannot be injected with a maggot from a fly then this meat would take longer to decay, as by bacteria. Let us say the longest period would be seven days. Another fact I would like to point out and this is by my own observation, a fly can not only lay eggs but it can Also lay a maggot. I will adjust this as I see necessary at a later date to fit into my time scale for the ten plagues. You will please note I am doing this as I go along editing. What we have now is three days without water and three days before the next plague. This was the plague of flies.

REPLY

This cannot be right because the fourth plague was of flies. So it must have been gnats or lice. All this talk about frogs is making me very jumpy. We now must remember that frogs are a water loving creature So it would not be long before they were feeling thirsty. This of course would affect their jumping ability. Now they would certainly be classed as hopping in our use of the word. They would try getting about on one leg. Over they would go and die. No sympathy please we knew that was going to happen. We now must do a bit of arithmetic. Three days without water. We now get the lice. They would not stay long on the dead bodies as with the heat the bodies would dry out quickly. As there are two scenarios we must give it six days.

NINE DAYS GONE

The third plague was of lice.

3. Lice (כִּנִּים): Ex. 8:16–19

The Third Plague: Moses, horned (a sign of his encounter with divinity), carries the rod, while Aaron, wearing the miter of a priest, stands behind him. The <u>gnats</u> arise en masse out of the dust from which they were made and attack Pharaoh, seated and crowned, and his retinue (by <u>William de Brailes</u>, collection <u>Walters Art Museum</u>)

"And the LORD said Stretch out thy rod, and smites the dust of the land, that it may become lice throughout all the land of Egypt." When Aaron stretched out his hand with the rod and struck the dust of the ground, lice came upon men and animals. All the dust throughout the land of Egypt became lice.

Exodus 8:16–17

The Hebrew noun כִּנִּים (*kinim*) could be translated as lice, gnats, or fleas.

We now have the reason why this was possible. All the dead frogs begin to rot. This can be worked out by the way nature works. We have the frogs body's rotting. I have gone back to nature to see what would have happened to all these rotting frog corpses. As Soon as the frogs die the lice would be the first living thing to leave as it lives on the blood of animals and humans. The lice after leaving the corpse of the frogs would be to search for food. These lice are not fussy eaters as my understanding of the lice population is there is a different strain of lice for different types of meat. There would be the wild reptiles and mammals. Remember I am using the zoologist's claim what would cause the death of the lice and what plague would follow? We would Also have the domestic animals of the human population. How long these would last with millions of lice sucking out their blood. I can just imagine the Egypt's trying to brush these lice of them I thing they are facing a losing battle you might as well lay back and let them enjoy their dinner. How long do you think this plague will last? Let us put it at three days as this seems to be a number that is wildly used. Their food supply dies and the maggots take over another three days pass and their food supply is extinguished. Here come the flies/gnats. It is a pity all the frogs died otherwise they would think Christmas had come early. I know Jesus' appearance was a long time after the ten plagues; this is just a figure of speech. That is why I said that Christmas had come early. I am Sorry but how long this would take I am stuck like a fly to flypaper. Total twelve days

The Fourth Plague: *The Plague of Flies* by James Jacques Joseph TisSot at the Jewish Museum, New York

The fourth plague of Egypt was of creatures capable of harming people and livestock. The Torah emphasizes that the *'arob* (עָרוֹב, meaning "mixture" or "swarm") only came against the Egyptians, and that it did not affect the Land of Goshen (where the Israelites lived). Pharaoh asked Moses to remove this plague and promised to allow the Israelites› freedom. However, after the plague was gone, the LORD «hardened Pharaoh›s heart», and he refused to keep his promise.

The word *'arob* has caused a difference of opinion among traditional interpreters. The root meaning is (ע.ר.ב), meaning a mixture - implying a diversity, array, or asSortment of harmful animals. While Jewish interpreters understand the plague as "wild animals" (most likely scorpions, venomous snakes, and other venomous arthropods and reptiles),

Gesenius along with many Christian interpreters understand the plague as a swarm of flies, If we take what happens in today's plague we now we have had a plague of mice in Australia. Locus swarms/plagues has affected all the continents the only place that has not been affected is the Antarctic. They will return practically every year up to twenty years. This is because they lay their larva which hatch out the following year. I will Also give this plague lasting three days as they have a life style the same as lice. Total eighteen days.

5. Diseased livestock (דֶּבֶר): Ex. 9:1–7

Disease Death rates during outbreaks were usually extremely high, approaching 100% in immunologically naïve populations. The disease was mainly spread by direct contact and by drinking contaminated water, although it could Also be transmitted by air. Initial symptoms include fever, loss of appetite, and nasal and eye discharges. Subsequently, irregular erosions appear in the mouth, the lining of the nose, and the genital tract acute diarrhoea, preceded by constipation, is Also a common feature. Most animals die six to twelve days after the onset of these clinical signs. This is what the LORD, the God of the Hebrews, says: Let my people go, So that they may worship me. If you refuse to let them go and continue to hold them back, the hand of the LORD will bring a terrible plague on your livestock in the field—and on your horses and donkeys and camels and on your cattle and sheep and goats.

We now have the plague on the livestock of the Egyptians. I have not heard or read any explanation by the archaeologists, zoologists or any other body that has explained the reasons for these happening; there has been none to explain the plague that hit the animals of Egypt, not to my knowledge. So here is my deduction. There are all these diseased frogs lying about and they covered all the land. This had to have an effect on the grasses that the animals were eating. It Also must have contaminated any hay that was harvested as food for these animals. This is my deduction for the plague on the live stock of the Egyptians. The time period must be six to twelve days. We will keep to what seems to be a usual time scale that is six days. We now have twenty four days.

—*Exodus 9:1–3* 6. Boils (שְׁחִין): Ex. 9:8–12[edit]

The Sixth Plague: Miniature out of the Toggenburg Bible (Switzerland) of 1411

Then the LORD said to Moses and Aaron, "Take handfuls of Soot from a furnace and have Moses toss it into the air in the presence of Pharaoh. It will become fine dust over the whole land of Egypt, and festering boils will break out on men and animals throughout the land." Boils occur when bacteria enter a hair follicle. Although the boil itself is not a genetic trait, you can be genetically inclined to be susceptible to the viruses. Boils start small and grow. Most boils are caused by staph bacteria.

They can be painful they can last from a couple of days to a week. The following gives another reason that is the cause of boils.

Lifecycle of *Leishmania*

Leishmaniasis is transmitted by the bite of infected female phlebotomine sand flies which can transmit the protozoa *Leishmania*. [2] The sand flies inject the infective stage, metacyclic promastigotes, during blood meals (1). Metacyclic promastigotes that reach the puncture wound are phagocytized by macrophages (2) and transform into amastigotes (3). Amastigotes multiply in infected cells and affect different tissues, depending in part on which *Leishmania* species is involved (4). These differing tissue specificities cause the differing clinical manifestations of the various forms of leishmaniasis. Sand flies become infected during blood meals on infected hosts when they ingest macrophages infected with amastigotes (5, 6). In the sand fly's midgut, the parasites differentiate into promastigotes (7), which multiply, differentiate into metacyclic promastigotes, and migrate to the proboscis (8).

The genomes of three *Leishmania* species (*L. major*, *L. infantum*, and *L. braziliensis*) have been sequenced, and this has provided much information about the biology of the parasite. For example, in *Leishmania*, protein-coding genes are understood to be organized as large polycistronic units in a head-to-head or tail-to-tail manner; RNA polymerase II transcribes long polycistronic messages in the absence of defined RNA pol II promoters, and *Leishmania* has unique features with respect to the regulation of gene expression in response to changes in the environment. The new knowledge from these studies may help identify new targets for urgently needed drugs and aid the development of vaccines as can be seen from the above animals and man when attacked by the sand fly will cause boils. The environment must have been just right to cause their larva to hatch out. And there death must have come quickly when the hail and thunder storms came. The LORD said you will must shelter from this storm or you will die. Let us give it another three days for the boils to really get a hold. And then the hail came, no more sand flies.

Up to here I have used my bit of knowledge about the rudiments of nature. But the following plagues do not seem to relate to the next plague. There are bodies that deal with the way weather patterns work and others that study the earth's contents and when thing happened, (Archaeologists.)

They have used this information to come to the decision that the start of the ten plagues was all started by a volcano erupting. Total eighteen days.
+—*Exodus 9:8–9*

7. Thunderstorm of hail (בָּרָד): Ex. 9:13–35[

This is what the LORD, the God of the Hebrews, says: Let my people go, So that they may worship me, or this time I will send the full force of my plagues against you and against your officials and your people, So you may know that there is no one like me in all the earth. For by now I could have stretched out my hand and struck you and your people with a plague that would have wiped you off the earth. But I have raised you up for this very purpose, that I might show you my power and that my name might be proclaimed in all the earth. You still set yourself against my people and will not let them go. Therefore, at this time tomorrow I will send the worst hailstorm that has ever fallen on Egypt, from the day it was founded till now. Give an order now to bring your livestock and everything you have in the field to a place of shelter, because the hail will fall on every man and animal that has not been brought in and is still out in the field, and they will die. The LORD sent thunder and hail, and lightning flashed down to the ground. So the LORD rained hail on the land of Egypt; hail fell and lightning flashed back and forth. It was the worst storm in all the land of Egypt since it had become a nation. It has been said that the hail weighed up to a hundred pounds each. (And I thought Manchester had bad weather.) There is no mention of another volcanic eruption and if there was I do not think the hailstones would be that big. We would must give another three days for this plague. Total. Twenty-one days.

Exodus 9:13–24

8. Locusts (אַרְבֶּה): Ex. 10:1–20

The Eighth Plague: The Plague of Locusts, illustration from the 1890 Holman Bible how anything can survive these extremely large hailstones I don't know. Maybe that being So large the livestock could dodge them. The locus had to come from Somewhere else besides Egypt. That's another thing that needs explaining by the biologists.

This is what the LORD, the God of the Jews, says: 'How long will you refuse to humble yourself before me? Let my people go, So that they may worship me. If you refuse to let them go, I will bring locusts into your country tomorrow. They will cover the face of the ground So that it cannot be seen. They will devour what little you have left after the hail, including every tree that is growing in your fields. They will fill your houses and those of all your officials and all the Egyptians—Something neither your fathers nor your forefathers has ever seen from the day they settled in this land till now. As we know when there has been a plague in our modern day of lotus they devour everything in their path. We must think. Where the plague of locus came from, before they arrived in Egypt. They must have been to a get-together at a banquet? I mean a couple of days before the country side had been devastated by a ferocious storm of hailstones weighing up to a hundred pounds. These lotuses had to be full as there was nothing much left to eat after the storm. That means that they could have got through Egypt in no time, but let us be on the generous side and say that they have had a belly full before they got there and didn't see the need to rush their afters. I think that three days would seem fit. That's twenty four days.

9. Darkness for three days (חֹשֶׁךְ): Ex. 10:21–29

Then the LORD said to Moses, "Stretch out your hand toward the sky So that darkness will spread over Egypt—darkness that can be felt." So Moses stretched out his hand toward the sky, and total darkness covered all Egypt for three days. No one could see anyone else or leave his place for three days. If this was me I think I would start believing that the Jewish god was god. I mean this is what I would call power cut. Maybe the Pharaoh was thinking that with all of the plagues that they had suffered the Egyptians had still managed to survive. They had suffered nine plagues maybe their gods were looking after them and what else could the Israelites god do that would make him change his mind. We know the plague of darkness lasted three days as the bible states this. The reason we can believe it was three days was the Egyptians had to go this long without a drink. Total twenty-seven days. We nearly have the grand total. After this, Pharaoh, furious, saddened, and afraid that he would be killed next, ordered the Israelites to leave, taking whatever, they wanted, and asking Moses to bless him in the name of the Lord. The Israelites did not hesitate, believing that Soon Pharaoh would once again change his mind, which he did; and at the end of that night Moses led them out of Egypt with "arms upraised". However, as the Jews left Egypt, the Pharaoh changed his mind again and sent his army after Moses' people. The Jews were trapped by the Red Sea. God split the sea, and the Jews were able to pass safely. As the Egyptian army descended on them, the sea closed before they could reach the Israelites.

Just one plague to go. *Exodus 10:21–23*10. Death of firstborn (מַכַּת בְּכוֹרוֹת): Ex. 11:1–12:36

Lamentations over the Death of the First-Born of Egypt by Charles Sprague Pearce (1877), SmithSonian American Art Museum.

This is what the LORD says: "About midnight I will go throughout Egypt. Every firstborn Son in Egypt will die, from the firstborn Son of Pharaoh, who sits on the throne, to the firstborn of the slave girl, who is at her hand mill, and all the firstborn of the cattle as well. There will be loud wailing throughout Egypt—worse than there has ever been or ever will be again." "On that same night I will pass through Egypt and strike down every firstborn of both people and animals, and I will bring judgment on all the Gods of Egypt. 'I am the LORD.'

Exodus 11:4–6
Before this final plague, God commanded Moses to inform all the Israelites to mark lamb's blood above their doors on every door in which case the LORD will pass over them and not "suffer the destroyer to come into your houses and smite you" (chapter 12, v. 23). Pharaoh realise that the Israelites population had increased So much that they outnumbered the Egyptians army and they could be a threat to his reign, with the threat of the tenth plagues he thought this would be the best time to give the Israelites their freedom. to follow the deities that God proved false. Exodus 9:15–16 (JPS Tanakh) portrays Yahweh explaining why he did not accomplish the freedom of the Israelites immediately: "I could have stretched forth my hand and stricken you [Pharaoh] and your people with pestilence, and you would have been effaced from the earth. Nevertheless I have spared you for this purpose: in order to show you my power and in order that his fame may reSound throughout the world." here

Pharaoh's taunt, "Who *[is]* the LORD, that I should obey his voice to let Israel go?» and to indelibly impress the Israelites with God›s power as an object lesSon for all time, which was Also meant to become known "throughout the world".

According to the Book of Exodus, God hardened Pharaoh's heart So he would be strong enough to persist in his unwillingness to release the people, So that God could manifest his great power and cause his power to be declared among the nations, So that other people would discuss it for generations afterward. In this view, the plagues were punishment for the Egyptians' long abuse of the Israelites, as well as proof that the Gods of Egypt were false and powerless. If God triumphed over the Gods of Egypt, a world power at that time, then the people of God would be strengthened in their faith, although they were a small people, and would not be tempted

Biblical narrative

The plagues seemed to affect "all the land of Egypt", but the children of Israel were unaffected.

For the last plague, the Torah indicates that they were only spared from the final plague by sacrificing the Paschal lamb, marking their place directly above their doors with the lamb's blood, and hastily eating the roasted sacrifice together with unleavened bread (now known as Matzoh) which they took from their ovens in haste, as they made ready for the Exodus. The Torah describes God as actually passing through Egypt to kill all firstborn children and cattle, but passing over (hence "Passover") houses which have the sign of lambs' blood on the doorpost. It is debated whether it was actually God who came through the streets or one of his angels. Some Also think it may be the Holy Spirit. It is most commonly known as the "Angel of Death". The night of this plague, Pharaoh finally relents and sends the Israelites away under their terms.

After the Israelites leave *en masse*, a departure known as The Exodus, God introduces himself by name and makes an exclusive covenant with the Israelites on the basis of this miraculous deliverance. The Ten Commandments encapsulate the terms of this covenant Joshua, the successor to Moses, reminds the people of their deliverance through the plagues. According to 1 Samuel, the Philistines Also knew of the plagues and feared their author. Later, the psalmist sang of these events. The Torah Also relates God's instructions to Moses that the exodus of the

Israelites from Egypt must be celebrated early on the holiday of PasSover (*Pesaḥ* פסח); the rituals observed on PasSover recall the events surrounding the exodus from Egypt. The Torah additionally cites God's sparing of the Israelite firstborn as a rationale for the commandment of the redemption of the firstborn. This event is Also commemorated by the Fast of the Firstborn on the day preceding PasSover but which is traditionally not observed because a siyum celebration is held which obviates the need for a fast.

It seems that the celebration of PasSover waned from time to time, since other biblical books provide references to revival of the holiday. For example, it was reinstated by Joshua at Gilgal, and, after the return from the captivity, by Ezra. By the time of the Second Temple it was firmly established in Israel.

We must take notice concerning the time of the exodus. And the ten plagues. The Old Testament Israelites were not "Jews." They were Israelites, but of no specific ethnic origin. The Bible tells us the people in the Exodus from Egypt were a "mixed multitude" (Exodus 12:38). The reason that we can believe this is because the Roman slaves came from all over Europe. There is a link that there were black slaves from Africa. (I mean to say it's not far from the Roman Empire especially by chariot.) This of course would be in line with the tribes DNA matching Jewish priests.

Plague 2 — frogs
Any blight on the water that killed fish Also would have caused frogs to leave the river and probably die.

Plagues 3 and 4 — biting insects and wild animals
The lack of frogs in the river would have let insect populations, normally kept in check by the frogs, increase massively. The rotting corpses of fish and frogs would have attracted significantly more insects to the areas near the Nile.

Plagues 5 and 6 — livestock disease and boils
There are biting flies in the region which transmit livestock diseases; a sudden increase in their number could spark epizootics.

Plague 7 — fiery hail
Volcanic eruption, resulting in showers of rock and fire.

Plague 8 — locusts

According to the UN Food and Agricultural Organization (FAO), when they get hungry, a one-ton horde of locusts can eat the same amount of food in one day as 2,500 humans can.

Plague 9 — darkness

The immediate cause of this plague is theorized to be the "hamsin", a South or Southwest wind charged with sand and dust, which blows about the spring equinox and at times produces darkness rivalling that of the worst London fogs.

Plague 10 — death of the firstborn

If the last plague indeed selectively tended to affect the firstborn, it could be due to food polluted during the time of darkness, either by locusts or by the black mold Cladosporium. When people emerged after the darkness, the firstborn would be given priority, as was usual, and would consequently be more likely to be affected by any toxin or disease carried by the food. Meanwhile, the Israelites ate food prepared and eaten very quickly which would have made it less likely to be contaminated. [However, this does not explain how the firstborn cattle alone Also would have perished.

A volcanic eruption did occur in antiquity and could have caused Some of the plagues if it occurred at the right time. The eruption of the Thera volcano was 1,050 kilometres (650 mi) away from the northwest part of Egypt. Controversially dated to about 1628 BC, this eruption is one of the largest on record, rivalling that of Tambora, which resulted in 1816's Year without a summer. The enormous global impact of this eruption has been recorded in an ash layer deposit found in the Nile delta, tree ring frost scars in the bristlecone pines of the western United States, and a layer of ash in the Greenland ice caps, all dated to the same time and with the same chemical fingerprint as the ash from Thera.[However, all estimates of the date of this eruption are hundreds of years before the Exodus is believed to have taken place; thus the eruption can only have caused Some of the plagues if one or other of the dates is wrong, or if the plagues did not actually immediately precede the Exodus.

Following the assumption that at least Some of the details are accurately reported, many modern Jews believe that Some of the plagues were indeed natural disasters, but argue for the fact that, since they followed one another with such uncommon rapidity, "God's hand was behind them". Indeed, several biblical commentators (Nachmanides and, more recently, Rabbi Yaakov Kamenetzky) have pointed out that, for the plagues to be a real test of faith, they had to contain an element leading to religious doubt.

In his book *The Plagues of Egypt: Archaeology, History, and Science Look at the Bible*, Siro Igino Trevisanato explores the theory that the plagues were initially caused by the Santorini eruption in Greece. His hypothesis considers a two-stage eruption over a time of a bit less than two years. His studies place the first eruption in 1602 BC, when volcanic ash taints the Nile, causing the first plague and forming a catalyst for many of the subsequent plagues. In 1600 BC, the plume of a Santorini eruption caused the ninth plague, the days of darkness. Trevisanato hypothesizes that the Egyptians (at that time under the occupation of HykSos), reSorted to human sacrifice in an attempt to appease the Gods, for they had viewed the ninth plague as a precurSor to more. This human sacrifice became known as the tenth plague. I had edited three pages of different views from different religious sects. I could have had millions if I had asked anyone with an interest in the bible, whether it was those that agreed with the bible or those that don't, but I am trying to keep the this book for the ones that do not have much knowledge about space and the bible. I hope I have not put in too much detail concerning the bible or space. Those that need to study these subjects more professionally; there is a wide range of avenues you can explore.

The Plagues of Egypt (Hebrew: מכות מצרים, *Makot Mitzrayim*), Also called the ten biblical plagues, were ten calamities that, according to the biblical Book of Exodus, Yahweh inflicted upon Egypt to persuade the Pharaoh to release the ill-treated Israelites from slavery. Pharaoh capitulated after the tenth plague, triggering the Exodus of the Hebrew people.

The plagues served to contrast the power of the God of Israel with the Egyptian Gods, invalidating them. Some commentators have associated several of the plagues with judgment on specific Gods, associated with the Nile, fertility and natural phenomena. According to Exodus 12:12, all the Gods of Egypt would be judged through the tenth and final plague: "On that same night I will pass through Egypt and strike down every firstborn of both people and animals, and I will bring judgment on all the Gods of

Egypt. I am the LORD." The reason for the plagues appears to be twofold: to answer Pharaoh's taunt, "suffer the destroyer to come into your houses and smite you" (chapter 12, v. 23).

After this, Pharaoh, furious, saddened, and afraid that he would be killed next, ordered the Israelites to leave, taking whatever they wanted, and asking Moses to bless him in the name of the Lord. The Israelites did not hesitate, believing that Soon Pharaoh would once again change his mind, which he did; and at the end of that night Moses led them out of Egypt with "arms upraised

The Israelites travelled the desert for forty years. Moses never got to the Promised Land. But after his death the lord showed Moses the Promised Land.

THE TEN PLAGUES

The grain growing in the fields would have died. That leaves the grain in the kilns. The scientist says that owing to the darkness the grain got a fungus disease. The grain is in kilns they do not see the daylight the darkness lasted for three days this was stated in the bible. The population of Egypt could not see an inch in front of them. That means by my knowledge of human life is that it cannot go over three day's with-out water. By my calculations it lasted three days. Who's a clever boy then? If the days and nights were So cold that it affected the grain than the human being would have succumbed to the cold if they were too frightened to move. The Jewish population sacrificed the sacred lamb. This, they then used as food. This then made them prepared for the next plagued. The death of every first born. They put a sign on the homes So that the LORD would know that there were Israelites in resident.

We know that in nature if there was a shortest of food the youngest would be the first to die. Any food that was available the oldest would the strongest and So it would survive. Soon after this last plague the Jews had to go into the desert So they must have had their own grain. The bible says that a poiSoned mist covered the land of Egypt that only affected the Egyptians. How did this appear and how could the previous plague cause this to happen. We can all see that the previous plagues were the course of the next plague.

With every one of the ten plagues all added up it lasted 3/4 months. I would like to know how anyone could have predicted this. With all the technology that they mustday and the anthologists have explained how it is possible that after each plagues caused the next plague.

REPLY

The boffins in all the categories dealing with the bible and the ten plagues seem to have come up with natural courses could have been responsible for these occurrences. They go into detail to prove their point. These plagues must have occurred over a certain time lapse. As this had not occurred before then no one has anything to go by. Moses must have had Someone to tell him of the sequences for these occurrences. The experts say that each plague would precede the last plague by natural forces. How could it be possible for anyone to predict these ten disasters? There are points put forward that a volcanic eruption was the start of the plagues. The nearest volcano was hundreds of miles from Egypt. So, Moses saw the smoke from the volcano and thought just what I need to start of the ten plagues. (Who's a clever boy then?)

The pharaoh gives the Israelites their freedom. He Soon regretted this decision. With all their labour gone the way of life of the Egyptians had changed. He called his centurion's together to go after the Israelites.

The Israelites came to this river and we must surmise that this had to be at a time when the tide was out. If they arrived when the tide was in there could be anything up to twelve hours before the next tide recedes. Remember it was quite a long time after the exodus from Egypt that the Pharaoh changed his mind. Let us presume that when they arrived at the point where there was no way that they could return to find another way across. The tide would recede gradually over the next twelve hours. And it would return over the same time scale. There must have been a lot of Israelites and this would include babies, children, old people and the livestock. If we recall the pharaoh was worried how the Israelites population had increased. We now have the Egyptians arriving

Do not forget the Egyptians probably had the comfort of chariots. We therefore must take it that the Israelites had got to the other side well in advance of the Egyptian's arrival. The time it would take them to cross over this land would have been a lot quicker. The position of this crossing had to be in a part of the river that was wide and deep enough to drown all the Egyptian's.

Bang Goes That Theory

It had to be wide and long enough So that all the Egyptians were in the river at the same time. The length of this crossing had to be of considerable length otherwise all the Egyptian's would not have been drowned. The scientists say that this was caused by the sea level dropping and a high ridge under the sea became exposed. It was a good job that this ridge was where it was otherwise the Israelites could have found their selves in a lot of trouble. This is a logic way of looking at things. The tide starts to go out and there is a large quantity of water trapped on the other side of the ridge. This would give the fleeing Israelites ample time to cross. For the sea to engulf the Egyptians there is two scenarios. One is that the LORD removed the ridge or he moved the trapped sea to overflow the ridge. This idea concerning the ridge is quite a reasonable explanation. If you look at all of the beaches you can see that the tide has a habit of coming in behind you in deep gullies. And before you realise it you are trapped. Everything in the bible can be explained as a cause of nature. The reason is that our LORD uses natural resources. This being the case, then after each miracle or disaster it must leave proof of these happenings. With our advancement in technology these can be found in our Earth's crust.

*Do you believe that D.N.A. is the most advanced di*scovery that tells us that we can find out who are relatives is? Do you believe in evolution? One is due to the scientist's experiments. The other is due to observation. I do not think that anyone can dispute these two scenarios. Because we have evolution this does not mean that there is no creator. Let us look at how evolution works. This is not a miracle. What we will do is look at us humans and evolutions. We have the advantage over animals because of our brain and this applies to our innovations. When these innovations become public knowledge, we are shown these and So we copy what we have seen. This applies to the animal world. They copy their parents they do not have the ability/brain use this to do anything that will improve their life standards. They might evolve if by accidentally they do Something and find it beneficial then they repeat it. With humans invading their territory they observe from a distance and were they observe us doing Something that they can relate to it they will copy it.

From Africa and proof of the ark

D.N.A. was taken from the eight priests. And when compared to the priests in Israel four of them matched. There are six families belonging to humans. These consist of. The original family line being. Adam and Eve. Then there was Cain and Abel. As we know Cain slayed Abel. That left one male. There were five daughters born to Adam and Eve this meant there was six siblings, So six D.N.A. lines. Let's Sort out where the partners came from. Cain was banned into the wilderness and got himself a wife. The five sisters Also got partners. 'Where from you may ask?' This is commonly asked question asked by those who do not believe in our God. To the rational people this seems to be the only way to look at things. Think! As you can see in the above comments about D.N.A there are six families. Here you will find where the other male and five females came from. God created partners for the female siblings of Adam and Eve. Let's take one scenario at a time. Where did a wife come from for Cain? This was when God laid Cain to rest. (I don't mean for ever.) This was when God most likely took a rib from Cain and created a woman. Now we are left with five women. There was only one thing that God could do that was to repeat his miracles. From the Earth's dust he created five males. That was the start of the human race. They came from the Soil; this is why you find all the Earth's minerals in the human body. So that is where the other humans originated from. Adam came from dust, and one woman from the rib of Adam. Man had one rib less than a female but here's where the evolving of nature takes place in the human body, there are many formations of humans having different shaped ribs and the number they have. You will find more about D.N.A, later, this will wake you up to D.N.A findings.

The only mathematical equations used are what I think are necessary. I will use them and then I will give my reason why they are not possible, or the reason that I dispute them. Also where I think that a theory is wrong I will give the logic for this decision. Sorry all theories are wrong. I will give my logic reason for the theory.

The physicists say that is no final point where Something ends. This depends on what we are talking about. Let's look at the smallest particle, the largest particle, the heaviest, or the lightest. These are Solids. There are places in the universe that will blow your mind that is if you believe in what the scientists and the physicists say. Let's take the most extreme case

The sun

So what coursed the Big Bang? There is only one explanation Something gradually caused a change to the structure of the sun, what can that possibly is? As we know the sun is hot but all around it is space. And there we have it, when you have two opposites i.e. the heat of the sun and a temperature at absolute zero one will be stronger than the other. This means that one of them must succumb to the strongest. Now which one do you think will get the upper-hand, of course the cold. How have I come to this as the reason? Our sun contains all the stars, planets, asteroids, comets and gases that make up the Solar system. My that's big. But let's think, the size of the sun has no comparison to the amount of space there is that goes on forever. As the sun's rays expand out into space, the cold atmosphere diminishes the sun's heat. But there is Something stronger than the sun and the further it travels into space the stronger it gets. I don't know if I am using physics to come to this decision or if I have just come up with a theory. What happens to the sun itself? There is only one thing and that is it starts to cool down. That means that in time a crust would form over the surface. And this would continue over billions of years. The crust would get thicker until the time would come when the heat would build up. This would cause Earth quakes, and in time volcanoes. As these spewed out there plasma this would harden which in affect would increase the thickness of the crust and So increase the pressure. When the crust became that thick and the pressure's rising Something must give. Wait for it. The sun gives in and surrenders to the pressure. *Bang*. The sun explodes, and the plasma spews out into space. This is happening to all suns. And the planets will Also come to this stage later Earth will be showered by fire and brim stone. This is Gods forecast for the Earth, don't worry he has Also said that he will intervene. You think this is farfetched. The bible told us this,

We have been told about the destruction of the five cities. that included Sodom and Gomorrah. These have been discovered in an archaeologist dig. They have found the places where these catastrophes have happened. With our advanced Technology that we have at our finger-tips it's amazing how many things that they say occurred in the bible are proving that these things did occur. There are a few things that they will not find and that are made from decomposing material, as in wood. One example would be Noah's Ark. If you think about it Noah and his family would have used the timber to build homes. Other items that would come under this category would be the chariots and their armoury of the Egyptians army. These would of course rust away.

It all goes hurtling into space. There are only two scenarios that could happen when this occurrence takes place. The first is the flying objects continue in a straight line this means they do not spin. As they are travelling at seventeen thousand miles per second and the sun's spin at its surface would not be strong enough to hold them into its orbit. If it did what you would have is millions of sun's flying around the sun had no gravity, and if it did when it exploded the gravity would have ceased. And all the matter would have got blasted out into space and there would be no gravity to contain it. It would have gone on forever, and that would have been the last of any Solar system and no universe. I am no scientist or physicist. So all this matter is going out into space there had to be Something to stop it. We know there must be Something. How or what caused this phenomenon. There is nothing I can think about could have caused it, only a creator. Now we have gravity. The first galaxies were created. In these galaxies there would be planets and meteorites, gases, and winds. I have just had a thought about winds. I have done what I always do and that is to come back to our galaxy. Whatever gives the Earth winds would Also apply to what accrues in space. The astrologers say that this is because of the moons gravitational pull. Phooey. I have thought about a new word for gravity and that's penny force. The reason is that the way the word gravity is used does not fit into my description of gravity? Gravity's definition as described by the physicists and the scientists means a force from a planet, sun or black hole that draws things in and this happens because of its spinning. The greatest spinners in the universe are our astrologers, physiatrists and all those that keep coming up with theories of how the universe works. the sun's. And not black hole as the physicists lead us to believe. The gases and winds would be at the outer edges of the Solar system. The larger objects would be held into the sun's orbit. The next scenario is they all had a spin. I worked this out for myself I bet you have as well. Now the astrologers say that Some of the planets don't have a spin. To me this does not make sense. It must be one or the other. Now lets be really ridicules. How about a God Ha-ha. Ha. I am falling over myself with laughter. Okay, okay. Let's use that scenario.

Bang Goes That Theory

I mean we must be fair and give everyone a chance. Hear we go. We will put ourselves in Gods place. Our sun has just blown up. I have created a sun and blow me down talk about paddy's law. So, you have a good look to see what you can do about it. All your hard work is being scattered into space. You must think of a way to stop it. Let's stop everything now. You take another look at your handy work. Well I will come back to that later. About one million years later you return. Everything is the same as when you left it. That's a disappointment you think, maybe I should leave it for another million years. I suppose I could start again from the beginning. No that would mean I am back to square one. You are now losing your cool in a spate of madness you send a force that collides with everything that has exploded. And low and behold everything starts spinning. I am glad I worked that out you say. I think I will leave that as it is I mean why spoil things when they are going well. Billions and billions of years later you return. You do not know how long it was I mean I would certainly need a long calendar. You take a look at your handy work and think who's a clever boy then. Before you got this brilliant idea, Some of the earlier planet and sun's were not affected by the force you had produced. I think I will leave them as they are, if ever I come up with Something else say a human it will give him Something to think about. Off you pop for another rest. A week later you return, you were going to take a bit longer but too much rest is no good for you. You start thinking again, we need Something else, Something different. You gather Some of the molten plasma and asteroids and clump them all together. That's a start, what is missing. I know a splattering of water that will cool the place down. It's a bit dark let's have Some light. There that's better as you look at a new sun you have created. Hello it's rather dark on the other side.

What can I d How about a lantern? That looks nice you are admiring your work when a gust of wind blows it out. That wasn't much of a successive you think. I know you gather more asteroids and low and behold the rays of the sun light it up. I am getting good at this you say to yourself. I think I will let things ride and return Soon. As everybody knows as you get older you are inclined to forget things. So Soon turns out to be later. Everything seems to have settled down, a sea had formed, and the water is cascading down the mountains. It still needs Something else. We need Some colour and movement. Got it you look down on your handy work and you are quite pleased to see a beautiful coloured tortoise moving along at a cracking pace. After two minutes you get a bit bored. Let's get rid of you and see what else we can do. Got it I will create the smallest creatures and the largest ones and that was the beginning. Bacteria and dinosaurs. I think I have mentioned my concern why the cave men have no drawings of rainbows; it has just come to me why there are no drawings of dinosaurs. That's me done for a week. There was only six days to a week, so you thought this was a good time for a rest. You take a day off and, on your return, remember to create Some plants well we all know you must have vegetarians, otherwise everything would eat everything else and there would be nothing left. Now you know how it all started.

How sun's and planets are affected by zero temperatures. This affects the circumference. This had the effect of Solidifying the molten laver as the crust got thicker So in time practically all the surface of the sun became a Solid. There would be pockets that would create pockets of volcanoes and Earth quakes. These worked like a valve to keep the sun from exploding over a longer period these Also would start to close So creating a pressure cooker effect. As the pressure built up, the sun exploded, that was when the Big Bang accrued. Now let's try a small experiment. When your fridge starts to get ice on the inside, this usually appears on the sides and the top. If you spray it with some hot water it changes the ice crystals into ice as you will see it is as hard as rock. It does not melt the ice crystals. If you keep applying hot water all that happens is that most of the water runs off and it does not affect the ice. There are places on the Earth below ground within a metre of each other were one compartment is well below freezing and right next to it, it is well above boiling point This Also happens where hot gushes occurs If you notice when a film has been taken, of an Earth-quake taking place below the sea, the heat can virtually stop at a short distance from its Source. Why does this happen the plasma flow is at the same temperature, but the sea keeps it contained, where the two temperature points meet. And then the heat stops. But the cold hardens the plasma. This is what happens when the sun's are cooling down. Also you will notice that there are fishes that seem to feed on Something that is right at the edge of this meeting point.

As the cameras do not zoom into this, we know that there is bound to be Some form of life that the fish can live on What does seem apparent is the heat stops and does not extend any further into the surrounding area. Why is this, it is having the same effect as the universe the cold is stronger than the heat. This then is what is happening to the sun's condition.

All the matter from the sun's exploding travelled in all directions. There would be Solid material, gases and molten plasma.

The planets/sun's would in time have the appearance of the first sun, they would in time cool down and then explode, So creating more sun's, comets and Asteroids. Now we have the question? What caused the universe to spin? The following brings gravity force controlling our sun's, planets and asteroids. If you have been taking notice my thoughts on gravity is does not control our universe.

Before the Big Bang happened, the magnetic field of the sun caused the universe to spin and this would Also apply to the gravitational pull.

All the sun's would have gravitational pull.

This gave the planets there gravitational field. This gravity would travel into space. At the same time the sun's rays travelled out into space. Does this mean that gravity travels at the same speed as light? I think So. The sun's rays and gravity have been travelling through outer space since its creation. Does this not tell you Something about these two wonders that the sun has produced? When I was younger we used to use old syrup, tin put holes all around the sides and light wood inside it. of the wood left the can. inter warmers

This means that they were all going away and being caught in the sun's gravity they were circling the sun. The scientists/physicists say the universe is getting bigger as everything is speeding away from each other. It Also means that they are getting further and further away from each other.

Now we have the question will these outside comets leave our universe or still be drawn into the spinning of the universe. Let us think about what we have learned from the previous logics. After each explosion of a sun the asteroids and sun's are getting smaller with each bang. As we know from earlier, materiel cannot increase So the sun's/asteroids are getting smaller.

The first sun would be in a void, there was nothing else. The sun's rays would be travelling through space. There was no need for any gravity, so this sun's magnetic field had nothing to apply its force to. The sun's rays have travelled through space since the first sun was created. This means that the sun's rays were invisible.

Space must have been a void of anything, no material no fluids, no gases

From the beginning of time there was space. All spaces temperature was at absolute zero. Then a sun appeared from nowhere. That's helpful. Because this accrued I am here, and I can tell this story. When the physicists or the scientist can explain this to me I will stop believing in the lord. That more than ever convinces me that there is a God (I am still a believer.) So, we have a sun the sun which has been the start of everything and with us for billions of years. This I will explain later why the physicists are wrong on three of their theories. What about the theory that the sun got hotter and hotter until it finally exploded. They can't have it both ways let's take a *cooler* look at this. The sun is hot but they now say our sun is cooling down. That to me means the original sun Also cooled down. Now let's look at these happenings billions of years ago. The sun's rays are invisible as is gravity. This can only be seen or detected when it has Something to react to. We know that there is a gravitation affect other-wise all loose things on a planet would not be able to stay on the planet. This sun was then omitting a gravitational force and the rays where expanding outwards. When it exploded it no longer carried on doing this?

The minute it exploded other asteroids, planets and sun's appeared in this space. The sun that exploded left ninety five percent of itself behind. Each of these sun's caused a gravity and sun rays and each sun was the centre of its galaxy with planets, asteroids, dust particles and gases circling it.

As for the sun's rays from each galaxy they would not pass into the next galaxy they would compensate and remain at the same brightness. Using this as a basic then I would surmise the same thing would apply to gravity. The sun's etcetera is moving away but remember that the light is travelling at a much greater speed, So it would travel into the space that was caused by the bang but as there is no substances/gases it would not show. Why would the sun's cool down? The sun's mass would be the same it would not expand other-wise all the sun's after the Big Bang would get bigger. In fact, the original sun and our sun would be expanding. Now let's look at the temperature of space. This is absolute zero. We know the sun cannot heat up space, So the sun must cool down. The reason for this is as I have said before as time passes by and the cooling of the sun continues let us see what would happen. There are hotspots on the sun as there are everywhere else. The Earth has hot and cold spots. Even in our garden. There would be parts of the sun where the molten plasma will cool. In time this plasma would turn into a crust. And this crust would get thicker. As this processed continued in time nearly all the sun's surface would be covered by a crust. The hotspots would be where the volcanoes are and faults would appear as the Earth is now with earthquakes. This would not end there it would still continue to cool. When the entire sun's surface has been covered by this crust, then the internal part that is the plasma, this would get hotter and hotter. Then it happened, the sun blew up. The Big Bang had accrued. This sun contained all the matter that is in the universe now. Let's go back and look at this scenario. Asteroids are not round like a planet. That means when the Big Bang accrued the crusts smaller pieces where the asteroids. The molten plasma was the sun's/planets. How have I come to this decision? The sun is the centre of the galaxy So there would be equal pressure from all sides this would cause the sun to explode in every direction. This would produce more sun's and more asteroids. These would all be travelling outwards at 17,000 miles per second. As they preceded on this course the sun's spin would pull all the planets into its orbit. The heavier planet would be the first to be caught in the Tilly Force. This is logical. The lightest ones would be the furthest from the sun.

spin on the first sun circling around it So the sun's would be the centre of each galaxy. This would give each a different gravitational field. Now we have a lot more sun's with a lot more comets and asteroids going around their sun. We have asteroids that are going around a planet each in their own galaxy. Talking about the spinning of the sun let me sort out the problem that the physics and scientists have and cannot work out. They have photos of the sun and they Also have instruments to show what is happening on the sun's surface. They have split the sun into three parts. One part for the North Pole. One part for the South Pole, and the middle part for the equator. They point out that the North Pole on the sun is spinning thirty times every twenty-six days. This is the same as the South Pole. The middle part, the equator is spinning at twenty-six times a day. They cannot see why this is so I know they have given these spins with a pinch of salts. These points can vary by such a difference at these given points. The reason they vary from the north to the South poles. Don't tell me not to be stupid everybody can Sort that out. Forgive me I am only new at this game. I will tell them why there is a difference. This is my way of explaining the difference about the way galaxies work. Spin a ball and the radius spins faster than it does at the top (North Pole) and the bottom, (the South Pole.) In fact there must be a point at the top and at the bottom when there is no movement. When I say this, I mean that even it was magnifies a million times I suppose you could not detect any movement but I am sure there must be Some.

Closer to home is the moon's surface, and close still is the Earth's surface. This is the reason why all impacts of meteorites and asteroids do not show a gorging on the surface but a direct hit. This was before Earth had its own atmosphere. This not happen in the beginning So what caused it. As with all other planets was it the sun's magnetic pull. Well let's look at the sun's that appeared after the first Big Bang. We know these are smaller sun's, So the logic is that the matter/planets from these will be smaller. I think before this happens the universe will look entirely different than it does now.

There is a theory that light travels faster than Einstein's theory of the speed of light. This is shown as a light travelling between two mirrors. With the scenario of two mirrors a beam of light passing from one to the other and returning. The two mirrors are on a train and the train is travelling at forty miles per hour, so light is travelling forty miles an hour faster than light Einstein's theory. The other demonstration is of a train travelling at forty miles an hour and a gun fires a bullet which leaves the gun at forty miles per hour. That say the physicists/mathematicians means the bullet/round is travelling at eighty miles per hour. We only need to use the guns scenario too prove the speed of light is going faster is wrong. Doing it this way is not as complicating as using the speed of light scenario. Let's take this in another dimension. A perSon fires a round in the opposite direction that the train is going in. a perSon on the station's platform see the perSon on the train fire the gun. That would mean that when the round left the train it would fall to the ground. This is how the person on the train should see it. This has been proved right So if I have printed any difference I am Sorry.

There is more on the speed of light and the effect of the sun's rays later that needs putting logically.

Every time a sun explodes the matter/particles get smaller and smaller. Saying that later an asteroid will hit the Earth and destroy it seems most unlikely. I guess God's prediction, that fire and brimstone will do that. And it looks more than likely now. This is what the effect of global warming will do. And why my prediction could be soon.

Global warming is occurring at the fastest pace than it ever has in the past.

(Too late now)

It is over the last ten years that global warming has started to really take an effect. The damaged has started if they stopped the burning of all fossil fuel now. The ozone layer will still be getting damaged. Not only do they must stop burning it now but they will need to find a way of reversing or if it doesn't repair its-self then we are in deep trouble. Let's take another scientific fact. We know the planets are moving further and further apart (more about this later) does this not convey the fact that now this makes more collisions most unlikely. And as far as destroying Earth surly the physicists and the scientists can see this. Now let's take the speed that they are moving through space.

They have proved that the planets in our Solar system are moving at 17.000 miles per second. This might not be correct to the last mile but I am not an expert, So I am not trying to educate anyone in the meaning of physics.

If we take the theory that a force is equal. A ton weight hitting the ground will move a ton by either compressing that amount of matter or it would splatter into little pieces. Something that produced the moon had to be the size of the moon or I think a lot bigger. If this was the case there must be an awful large crater Somewhere along earth's surface. And if it was that large, it would have destroyed the Earth. We'll all other planets have been hit by meteorites. And the force would be a lot stronger, as they have no ozone layer to protect them. To me that means that over billions of years all planets have been bombarded by objects and the universe should be one long desert.

All this space exploration is a waste of time and money. The money spent and being spent it would have been better spent on innovations on this planet. As this is the only way life will and has improved man's life style, and will be a help to man. They keep moving the goalposts back. When they discover another planet, and they find nothing they say they we will find it on the next planet. This is to keep them in a job. This is another way to waste taxpayer's money.

I was always bottom of my class in school, this was in all subjects. In fact, if there was a basement I would have been the only child in that class the optimistic view is I would be top of the class.

So if I am wrong with any theory. So, let it be. I am hoping most of my thoughts are based on logic.

Talking about thoughts what is a thought. I had an imaginary friend when I was a young child that I used to talk to. My mother use to say to me who are you talking to. I used to tell her, and she would shake her head. As I got older this phase stopped. I have worked it out in my later life that I am no different from most people. Then it started to happen again. I began answering the actors on the television. Also, I did it with the radio programs. It suddenly came to me that I was no different. Football fans are always calling out to football players when there is a match on the television. They shout out what they think about them or how they are playing. It is no different to what I do.

Everybody has thoughts. Thoughts are talking to yourself without the words.

And the lord said man shall live on the Earth, but the heavens belong to me. If you put this into logic it means that our Lord was telling us that there is no point in space exploration.

Does this open the eyes to the Big Bang and what is the logic of black holes, space and outer space?

Not a theory, logic. The universe is expanding but how about Something you might not know.

Black holes what is the logic of black holes?

Comets, meteorites, planets, gas, ice. These will Soon to be answered.

So, what coursed the Big Bang? There is only one explanation Something caused a change to the structure of the sun, what can that possibly be? As we know the sun is hot but all around it is space. And there we have it, when you have two opposites i.e. the heat of the sun and a temperature at absolute zero. One will be stronger than the other. After a while then, one gets the upper hand. Now which one do you think will get the upper-hand. It is of course absolutely zero. That means that the circumference of the sun which is in its plasma state would have the effect of Solidifying the molten laver. As the crust got thicker So in time practically all the surface of the sun became a Solid. There would be pockets that resistance that would remain as plasma but not for long. These pockets would of course be volcanoes and Earth quakes. These worked like a valve to keep the sun from exploding over a longer period these Also where closed So creating a pressure cooker effect. As the pressure built up, the sun exploded, that was when the Big Bang accrued. Now let's try a small experiment. When your fridge starts to get ice on the inside, this usually appears on the sides/top. If you spray it with Some hot water, it changes the ice crystals into ice as you will see it is as hard as rock. There are places on the Earth below ground within metre of each other where one compartment is well below freezing and right next to it, it is well above boiling point This Also happens where the hot spot gushes occur. All the matter from the sun exploded which travelled in all directions. There would be Solid material and mostly molten plasma. The small matter that was molten plasma would of course in no time be of a Solid material. These two would be drawn into the sun's orbit and So create an asteroid belt. The sun's would in time have the appearance of the first sun, they would in time cool down and then explode, So creating more comets and Asteroids. This has not yet happened otherwise we would have had a lot more galaxies. These of course would be a lot smaller. Now comes an unanswered question. What coursed the universe to spin? We know that as an object travels through space there is nothing there that can make it spin. It could not be the Big Bang as all the matter sped away from the bang, and there was no gravity. That means that they were all going away in a straight line. As I have pointed out when you have a problem let us put our two feet on the ground and look at it from an Earth's scenario. If in space nothing can course a spin how you can expect this when there would be a lot more reasons on Earth to cause a spin. If you drop Something it does not start to spin. What can we come up with to cause this? The scientists/

physicists say the universe is getting bigger as everything is speeding away from each other. It Also means that they are getting further and further away from each other. This should mean that the gravity pull must be getting weaker. I disagree with this. Now comes the question, will these outside comets leave our universe or still be drawn into the spinning of the universe. Let us think about what we have learned from the previous logics. After the Big Bang explosion, the asteroids, planets and comets are getting smaller with that Big Bang. As we know from earlier, materiel cannot increase So the sun's/asteroids are getting smaller. Light and Penny Force are travelling beyond our universe and not ending at the edge of it. The reason why we can not observe this is because there is nothing there. Light and gravity can't be seen when there is no matter for it to reflect the light or show any moving objects.

Let's do an experiment. When I was young, and we had bonfire night we used to have winter warmers. These were made from an empty tin. What we used was syrup tin. We made holes around the sides of the tin. We filled the tin with paper and wood. We then set it alight. When the fuel began to burn we swung it around. It caused it to get very hot. When spinning at top speed not only did the wood but the flames never left the winter warmer. This goes to show that when Something is spinning around the force it creates holds all that is in that force stays there. Now if we take this as a sun or a planet the heat stays near the sun but gases escape. Now these gases are trapped by the rotation of the universe. In time the contents of the tin will cool down. If the contents were of boiling liquid what would happen is a crust would gradually form. This is what happens to sun's. A planet was a small sun from a large sun exploding. Owing to its size it cooled down quicker. And So, we have planets.

As the physics know the winds are caused by the hot air meeting cold air. When they reach the land the mountains and valleys alter the winds speed So that it increases the speed and can change its direction. This Also applies to rivers and valleys. This might not seem of any importance but if you are thinking of going mountain climbing it could save you money in buying winter attire

Sorry they have it wrong again,

Lot of planets do not have a tilt; So they have no seasons they don't have any indirect movement only the forward movement caused by the

Big Bang. I think that these were the first planets from the first Big Bang. Also these planets managed to dodge all the objects flying around So no tilt. That is if you want to believe the physiatrists and astrologers. If you do not know my opinion about this then I must not have written about it. Sorry then I will mention it later. The Big Bang propelled them on an outward projection. The physics /scientists have put forward the scenario that another planet/ comet or meteorite hit the Earth; this then caused the tilting affect. It must have been a big object I mean A BIG object. If it was a glancing blow there would have been a long gorge for this to have any affects. And if this object made a glancing blow as it passed the Earth the moon would not be round.

The Milky Way was the original place of the first Big Bang. So that means all the matter was contained in that sun. (The physics/scientists have not explained where that sun comes from. (You can't get Something from nothing.) When the scientists can do this then would have convinced me that there is no creator. I don't know much about space only what I have seen on the television. Let's start at the beginning? If the sun started as an atom/nebulous and it started to multiply why did it stop, physics say it grew hotter and hotter. So when it exploded and it created the Milky Way why didn't the sun's it created continue to grow. As everything available to create the original sun must still be available to carry on that process. Also, you must think the idea that sun's got hotter goes against all the physiatrists logic. They and scientists say that the sun is getting colder. This is fact the true case of what happens to a sun after billions of years. In time to come the sun's in the universe will come to the end in the same way that happened to the original sun. The sun's created by the Big Bang will Also collapsed and explode and So the process will continue. After each explosion the sun's/comet and asteroids created would be considerable smaller and smaller.

So how an asteroid/comet hit can hit the Earth and caused it to tilt is beyond me. As these objects were getting smaller and smaller with every. When the original sun collapsed, the comets/asteroids that it created were smaller objects. Why are comets and asteroids still bombarding Earth? To remind you of the way that our Solar system works, and other galaxies works. As the Solar system expands objects on the outer edge travel faster. As the asteroids are the lightest they will be thrown out of their position in the Earth's orbit. I hate to admit it but this is a theory. I have just had another thought. Earth's position in the universe

Each sun and the Earth have their own magnetic force field. When seen from any position, as from Earth. The universe is moving in a clockwise direction. Now we can see that in most cases each galaxy is moving in a clock-wise direction. This means that where the edges of each galaxy meet they are colliding in the opposite direction. The sun's rays and the gravitational pull of each galaxy is balance out. One is not drawing the other into its galaxy. This means that there are areas where three, four or more galaxies meet. Would this give the appearance of a black hole, this would mean that all black holes would be in the same place. This does not seem right to me. My other explanation seems more reasonable. This is explained later.

No black holes

The sun's rays cancel each other out; as does the gravity how come that the experts in the universe have not come up to with this. It is the basic of how things work. This is Isaac Newton's discovery.

ProfesSor Cox shows a diagram on how space and time are connected I have taken a look at this and I have come to the decision that he is not correct in his way of looking at this sum.

The physicists. If the moon was controlling our seas how does it manage to move them in two different directions? Light comes down straight down? See below. Sound comes down straight So moons gravitational pull must come straight down, if there is such a thing? Gravity has a different force in different parts of the Earth. If this is So can anyone tell me how the moon can change its gravitational pull can do this? The physicists say that sun rays bend. If this was true then what stops them bending all the way around the Earth. My logic is that the particles in the Earth's atmosphere reflect the rays. Scientists have now discovered that in all Europe there are only, six family lines. So Europe families travelled to the far corners of the world. I would like to know how many different family lines there are in the rest of the world. I. E Africa and Asia. And do they originate from the same six families as this is my observation of the bible families. The DNA taken from all the other continents have arrived at six families. Why is the

Earth spinning on an axis? What coursed this as we are the only planet in the universe to have seasons? The lame reason is we were hit for a six by a passing object that didn't like the way were heading and decided to change all that. If this was the case there must be an awful large crater Somewhere. And if it was that large, it would have destroyed the Earth. We'll all other planets have been hit by meteorites. And the force would be a lot stronger, as they have no ozone layer to protect them.

The cost of space exploration is tremendous.

BENEFITS? PRACTICALLY ZERO.

All this space expiration is a waste of time and money. For money spent and being spent would have been better spent on innovations on this planet. As this is the only way life has and will be a help to man. They keep moving the goalposts back. When they discover another planet, and they find nothing they say they we will find it on the next planet. This is to keep them in a job. This is another way to waste tax payer's money. Our galaxy was created in a different way than the rest of the universe. The universe originated from the Big Bang. This was molten plasma containing all the metals and elements The Earth originally should have been like a sun then as it cooled a crust formed until there was one land mass. Is there any other moon in the universe that has as much molten plasma as the Earth? The continents are thousands of miles apart. The seas where supposed to have eroded away the land, they have done a magnificent/miracle job to chip away So accurately that all the different continents thousands of miles away slot together. I will tell you later why this is not possible. Over the globe than Earth must have been all land. So where did the water come from? If all the land was one mass and now the continents are positioned all find one mass and then insert I was always bottom of my class in school, this was in all subjects. If I am wrong with my theory. Let it be logical

When an astronaut leaves the space capsule in space in which direction does he go? If all the planets and stars are moving away from the sun So does he. As gases are dispersed in the universe they are at the edge of each galaxy. The astronaut being lighter Also moves away from the space ship.

This would mean that in a billion of years' time he would finish up just beyond the asteroid belt and the gases. The scientist is going to have the latest space ships spinning, this will course will give the space ship to have its own gravity. Gravitation is caused by the planets spinning. That's why as Soon as you leave the Earth's atmosphere there is no gravitation pull as you are in a vacuum as with all the other planets.

The sun consists of liquid /molten plasma if this is the case and it exploded, then all this liquid would be dispersed into the vacuum they would not be concentrated, as a planet or a sun. If an explosion accrues in the water the bigger the explosion the smaller the particles of water will be thrown out. How then did they form into sun's/planets?

If the winds on the Earth are usually up to 160mls per hour then, If everything has evolved from the sea i.e. from bacteria, and we are the most evolved then why is there still bacteria. And what did bacteria evolve from. They say they have just found life form in the atmosphere and it's too heavy to rise from the Earth. It has come from outer space. Metro 12/09/13? If it's So heavy why hasn't gravity made it come down to Earth? Also meteorites get burned up coming through the atmosphere. So this form of life must have drifted down. The scientists should study these as they can defy gravity. I don't know how they work these things out. As they say its theory like the one about the teapot in between the moon and Earth. My theory is they are flying pigs. Which would prove that pigs do fly?

The moon. The scientist have discovered that the centre of the Earth is Solid, they do not say why the centre is Solid well here could be the answer. (Read on to find out why) I can predict that all planet and sun's have a Solid center. How do I come to this conclusion? The scientists, physicists do not know if the moon is Solid or the core is molten plasma as the Earth. The moon is a fascinating object in the sky. One reason being it is one of the largest natural satellites in the universe. This is mind blowing. This is putting the cart before the horse. Let's skip back, I mean really go back, no further back, now you have it. In the beginning God created the sun, to light up the Earth. And he created the moon to light up the night. If this is So, then the moon was as the sun was a star. It being of a smaller size and not having the ozone to protect it cooled down quicker. We can deduct from that that the center is of molten plasma, the core is a Solid. I will tell you why later.

Bang Goes That Theory

There are pictures taken of the moon and their craters, it shows flashes of dust rising from the craters. They say it is caused by objects hitting the moon. I need to know if all these light disturbances are in an original crater. We know that there is an atmosphere on the moon and gravity is not as strong as the Earth's. If there is an atmosphere why is it we can not see these objects going through the atmosphere? we can see the moon landings of our space ship. Is it at all possible that these flashes of light are being produced by eruptions coming from the inside of the moon as with an explosion with the building up of heat? How long is it that they have been seeing these flashes of light. If this is the reason, then it will answer the question that the scientists and physicists want to know if the moon is a Solid or not.

I need to think this out. I will let you know later. Got it all sun's and planets have a Solid center this happens because of the cooling effect of the absolutely zero temperature. The heat in time will use up all the oxygen. And the center of the planet/stars plasma cools until it I a Solid as in metals. With being starved of oxygen it cools and then becomes a Solid. This has the effect as it gets bigger and the crust on the moon's surface gets thicker there is less heat being produced and So there is less heat being trapped. This is only a theory, but I hope it has a bit of logic to it as you might of surmised I am no scientists So maybe it will give them Something to think about. The moon is far advanced as in the cooling process. So in a few years' time will it still continue to cool from the inside and from the surface or will it be another bang in our galaxy. Before our Earth had an atmosphere there would be nothing to burn them up. The means craters, and volcano could be seen on our Earth's surface.

Moons eclipse. The scientists/physicists say that the moon becomes red from the reflection of the sun's rays being reflected from the Earth. It filters out the blue. If this is so, why then why do the rays when entering the Earth's atmosphere do not have the same effect? This should show to people on Earth as a red planet. When there is a complete eclipse all the moon's surface is red. They say that the reason for this is the light is reflected. When the moon is only part covered there is a part of the moon is white. This is the sunlight having a stronger effect than the red. I will give you my reason why the moon shows a red glow and Also a white one.

Q1

The astrologers and geologists say that the earth's land mass was one land mass; this has now been proven by satellite photos. And that all the continents fit together like a jigsaw. What I would like to know is. It has taken fifteen billion years for the seas to corrode away the land leaving the continents to still look like they have had no erosion. I'm no physicist scientist, mathematician or Geologist So over all that time (corrosion) has it manage to corrode the land to leave such an incredible contour that still interlocks. If this fact is So then the continents are still being eroded, but more land is being created on the other side to the erosion' that means to me that the continents are moving closer together in there north hemisphere.

How is this for a knockout blow that has not been seen by astrologists, scientists and physicists? The Earth's surface was partly covered by land. Surly if a planet or a sun was cooling down then it would not cool as one land but would be in patches, *like islands*. Where there are two forces such as heat and cold battling against each other the cold force will always win. So as in the scenario with a sun and the below zero temperature all the surface of the sun did not cool into a Solid, it cooled gradually as in patches. The appearance of a volcanic eruption, or a crater coursed by a comet will look different. They must be different. I would think that the sides of each would be able to see as a volcano sides would be higher. Also, a meteorite hitting the Earth would show a smoother ridge around the circumference.

The reason that that all things were found buried for a certain amount of time is not that they became extinct, they had moved with the changing of their environment. It is more noticeable now because it is moving faster. Because of global warming this is affecting plants, insects and all other forms of life, which took millions of years, you will notice that when I talk about how long ago I do not go into how many as it will confuse you, I am already confused by it. As we know the Big Bang was a sun exploding.

The sun had been there a long time before it exploded So all the metals in the sun have not been affected by radiation. The sun's radiation travels outward that must mean that the matter it spurted out had radiation in it that was billions of years old. Then the radiation from sun's rays gave the appearance that it was older. Scientists studying this radiation would automatically think that by studies that it was older. This does not apply. Why you might ask?

What does this tell you? It tells me that if we could replicate what the sun does with radiation then we would be able to reduce or even get radiation to a state that was not harmful.

Plants, insect, animals, and all other things can now be seen to be moving with the change in the environment.

Sun's gravity can make the planets e.c.t. go around it but the speed that they are moving away is more powerful.

Black holes

If this is the case then the black holes cannot draw anything into it.

Now let's look at gravity. The physicists, scientists say that the gravitation has the weakest strength of amongst all the other wonders of the universe. So which is it?

.((north/South moved= no ice age.)

Answer. (Read the bible)

Einstein's theory of relativity.

As for time in space changing the age of people, that the parents become younger than their children this is not logical. If the moon was controlling our seas how why don't the tides all move in one direction. Does it manage to move them in So many different directions? All the functions concerning the world's seas and winds are the spinning of the axis. And the different contours on the planet. Find out later what courses the tides to rise and fall. In the beginning there was one sun. If there was the Big Bang and there was one sun? And it exploded why is there still a sun my theory to follow. If these sun's in the universe collapse and then they explode, the sun must be the same compounds as planets as they derive from the sun So the same thing must be happening. The sun's/planets gradually cool down. This causes a crust to form, as it did on planets. So the original sun formed this crust which gradually got thicker which then heated up. There were probably sun quakes and sun volcanoes, until it became harder, for the sun to release the heat. So, when it exploded the crust disintegrated to form other planets and sun's. This means they are all like a pressure cooker and if they keep heating up without having the abilities to cool down they will explode.

Black holes particles.

The reason for black holes is it works opposite to the Earth's atmosphere; these are pockets in the vacuum where there are no particles in the atmosphere therefore there is no reflection of the light from the sun. That's why going beyond our universe into outer space nothing can be seen. Going by what the psychics tell us then all outer space is a black hole. So if they are right then all our Solar system are being drawn into it, the Solar system are not expanding they can't have it both ways. So let's Take my logic and agree with their first logic/observations, that the Solar system is expanding. The appearance that black holes are drawing in all matter because its gravity is So strong that not even sunlight can escape this is preposterous this does not make sense. The reason for this observation is that the planets are moving further apart, So on that presumption the black holes are Also expanding this gives the assumption that all matter around the black holes are being drawn in. I have only recently started watching programmes about space, and by what I have seen is when seeing all the Solar system there is blackness I don't know if this is true but if it is it bears out my logic that there is no particles in the outer limits. More later about black holes

Dinosaurs why have they not evolved

Why has the female/ human /animal body not evolved to give birth without having pain? I would have thought that would have been the first thing to do. Calcium is not found in nature So where did it come from. As all living creatures have bones, and teeth the ice age did not accrue. What they are implying does not make sense; this would bring us to the conclusion everything that was of a living nature died. These then fetchers us back to the beginning. If, as Some sects believe we came from outer space then this means that they have made another visit to are Earth, and the whole recycle would must start again. You might think I am being side tracked but the following has a connection to the previous observation about the ice age. The archaeologists have found when digging that there are iron based rock that appear to have moved. There are large pieces and they gradually reduce in size as they get further away from the North Pole. Continuing their exploration at a lower/higher level they have done an about turn. The explanation they give is that the poles have changed their position. Scientists understand that Earth's magnetic field has flipped its polarity many times over the millennia. In other words, if you were alive about 800,000 years ago, and facing what we call north with a magnetic compass in your hand, the needle would point to 'South.' This is because a magnetic compass is calibrated based on Earth's poles. The N-S markings of a compass would be 180 degrees wrong if the polarity of today's magnetic field were reversed. Many doomsday theorists have tried to take this natural geological occurrence and suggest it could lead to Earth's destruction. But would there be any dramatic effects? The answer, from the geologic and fossil records we have from hundreds of past magnetic polarity reversals, seems to be 'no.' to my way of reasoning things is not the magnetic fields that cause the iron deposits to seem that the poles have moved. The scientists have said that the poles have moved from north to South but have given no reason why this is so. Let me of average intelligence tell them what has caused this extraordinary happening. It is not possible for large lumps of iron to pass through the Earth's crust and the smaller ones would follow. Ipersonally would have thought it would have been the other way around. As they have said this occurrence took place over billions of years. Now let's look at the scientist's view that the metal elements that we have here on Earth came from a star exploding. The reason they give is that the iron in the star was brought together into the centre of this star which caused it to explode. They say that as Soon as the star produced iron it was sealing its death warrant. Let's take it beyond the iron and look at the other metals. These were Also being drawn into the centre of the star. The heaviest ones are nearer the centre and the other metals fall into place in order of their weight. I do not know the reaction you would get when the sun explodes. If the centre of the sun contains these metals, and the rest is the molten plasma, and this explodes in an outward direction from every part of the circumference would the metal stay in the same place. maybe the heat produced to create the bang melts the metals this enables it to disperse as molten plasma. What is reason why the metals in all the planets and sun's have a core? I have tried to read up on this but now I have not come across anything. The next information is based on my observations based on what makes metals go hard on Earth and take those finding and apply them to other scenarios in the universe. When the heat is removed the metals go hard. This cannot be applied to metals in a place where there is a constant heat force, so what happens. All the oxygen has been removed by the heat factor. So, if the is no oxygen then there is no fire and there is no heat. This is very confusing. Let us look at what all the metal experts are saying, all the metal that is in the centre of the sun's and the planets whether they are gas or of Solid a Solid mass the sun's produce their own metals. Over time these metals finish up in the core of sun's and planets. The reason I have come to this decision is because at the time of the Big Bang only a small amount was blasted into space that means that the centre of the sun's centre remained where it was. In time when our Earth has cooled to the point where it will explode the same thing will happen be it on a smaller scale. This is when Isaac Newton's equal forces apply. As this keeps occurring will there come a time when all that is left is atoms.

Gases do not make sun's

Take what happens when the explosions here on Earth takes place when a volcano erupts. The plasma spills out with all the metals in its contents. This occurrence is over billions of years, the in there exploration of the Earth's crust they have found that the iron does a u turn So that the largest ones point to the South followed by the smaller ones flipped its polarity many times over the millennia. None of them explain why. Let me save them a lot of time in making up theories. We know that the Earth's axis is spinning at 23 degrees, what could possibly happen to cause this tremendous turn around. The spinning of the axis as proven by scientists is 23. degrees. Now comes the crunch of this observation. It is not moving in a fixed direction to the Earth's circumference. It has a rocking effect; it is gradually moving towards the north. This has the effect that the north will finish up in the South and by my brilliant deduction the South will finish up in the north. Later I will give another reason why this is So

Is it only on the Earth that we have one of the poles that radiate magnetic fields that makes a compass point to the north?

There must be a difference the way metals and molten plasma reacts in an explosion.

Bang Goes That Theory

I am comparing the explosion where there is a gravitational pull and when it happens in space. On the Earth the heaviest metals carry further away.

If you have taken notice of my reason why stars explode you can see this process happening here on Earth.

Why do the poles move it is the Earth's 23degrees axis? This is more logical. Why was it coursed by the spinning of the world on its axis? The spinning is not on a fixed rotation, it may stay at the 23degrees level, but it spins at an irregular angle. This means that it gradually moves in a Southern direction. That is why the poles have moved. This then effects the weather So that there is a slow movement in the way the weather changes. The poles moved slowly to the opposite sides of the Earth. As this was a gradual movement the plants moved with the weather, As did the animal and insect population with this changing of the environment, the insects and animal followed their food supply. The experts say there has been more than one large

There was no ice age in the sense that the entire planet was frozen, So the dinosaurs would have moved with the change in their environment. The archaeologists can't understand why the dinosaurs died off. They know there was no shortage of food and they didn't die of diseases. There was no catastrophe from a gigantic explosion of a meteorite hitting the Earth, as all comets and meteorites are getting smaller as each star collapses. And as the Solar system moves further apart the chances of anything hitting anything else gets more remote. And if they did, the impact would be like a pea hitting the planet. Does anyone think that there was a great flood? Would this explain it, say it rained for forty days and forty nights. The vegetation would not be affected as being So deep in the water it would not be affected by oxygen, It would not have decomposed. Going by what the bible says there was six or seven animals that were brought on to the ark. There were six of every clean animal and two of every unclean animal. Let us look at this statement. There is nothing in the Ten Commandments that say there are clean and unclean animals. This is man bringing in stupid traditions. There are lots of different religions with their traditions that have nothing to do with the word of God or Christ's teachings. All these traditions are there to keep people that have not the strength to bow down to their wishes. Most of these people are the females and the children.

EVOLUTION

If its millions of years since the dinosaurs died out why haven't they evolved again? Why is it that we can see evolution in the passing of time in all animals, bugs, bacteria and with every living thing on the Earth and the remains are being dug up to prove this fact. But there are far too many missing links. Never mind about the missing link between man and apes, that's the easiest one to work out and they can't manage to do that. What about all the millions of missing links. Why it is only over the last five thousand years that man has left any evidence of his being here on Earth as an intelligent being. Which brings us to the evolution of women? Over the millions of years why haven't woman evolved So they do not feel any pain during child birth the animal world do not suffer in giving birth. The young when born must be able to stand and run in no time. This has not evolved over millions of years because before they had this ability they would have been horse meat in no time. The human species have not been able to evolve So that it was protected from anything that would protect them from predators. The worst thing that could befall a species is the smell of blood the female race would not have lasted five minutes if they had periods and if this came through evolution no species would evolve that put them in the position that they were more likely to be slain by their predators.

How could man survive the hazards of living in an environment were you were the easiest pray for all types of meat predators. This Also fetches up the problem what was the food that this species prayed on. They could not live on the plains as they would be the slowest thing only having two legs. That leaves the forests. This must mean that they lived in the trees. There not going to make a monkey out of me. I can't stand heights any way. And humans would not last long living in the trees. If we were descendants from the ape family at what time would we be able to evolve to feel safe to stand on our own two feet. And can you imagine of our species giving birth up a tree. The females would must come down to earth and be ready to shin back up a tree in no time. As far as us evolving you can see why I have a problem with humans evolving.

The archaeologists say that man has been on the Earth for millions of years, why has it taken So long to evolve to our civilised state. We see animals/birds evolving all the time around us. Most of them learn by their adults Also they have the basics as in avoiding danger, and which food they are supposed to eat. The food they are supposed to eat can change in a matter of days, this is due to the fact they will try other types of food. If they don't then they die. Most species die off because their food supply runs out. Thank god for God.

If any life was going to be created, then surly it would have occurred just after the first Big Bang and evolved from there. I mean to evolve from bacteria they would need the longest time that was available to reach from there to us humans.

When a black is photographed you cannot see what is behind them. This does not mean that the Black hole has devoured them, what it means is that there in nothing behind it.
Planet/comets and asteroids got smaller when the sun exploded. How can you see a black hole in the sun's centre with the strong light it produces? The Black Hole is there because the sun's rays are not being reflected. Here on earth when we have a sky of broken clouds, and the sun's rays shine through the nearest clouds will look like a Black Hole. That is because there is no sunlight to reflect the sun's rays.
How is our Earth situated in respect to the other Solar system? And how does its size compare with other sun's. If we use physics it should be the smallest sun in the universe. There is another question that comes to mind. As it is written that our Lord brought together the fire and rocks from the skies to produce the Earth and the moon. If this is So then the nearest galaxies would have less matter. Solar system are moving further away So they are unlikely to hit Earth and their effect would be like a pea

hitting the earth. In the beginning there was one sun where the hell did that come from, let's have a theory it started with a nucleus So why did it stop growing physics could say it got too big and blew up So all the sun's it created should be growing. Here is the question again. Where did the nucleus come from? The boffins stand back in amazement and say, I never thought of that.

The scientist is going to have the latest space ships spinning this will cause inside the ship to have its own gravity. Why does the moon not cause a gravity field? So, what causes the poles to change places over millions of years?

The end of the Earth as we know it

Here's Something to think about it is about the idea that we came from space. Another theory is that we evolved from bacteria from out of the sea. Then we have the Earth being flooded more than once. then comes the theories that we have had the Earth being covered in ice quite a few times. There're is Also the theory that the Earth was blanked out by a stupendous volcano that blanked out the sun and all things perished. This Also applies to the theory that the Earth was hit by a massive asteroid. Just the thought of this gives me a headache.

EINSTEIN'S THEORY OF RELATIVITY

As for time in space changing the age of people, So that apersons Parents become younger than their children, this is not logical. Where they have all got it wrong, is that distances change but time doesn't. If the winds on the Earth are usually up to 100 miles per hour then, if the moon was controlling our seas, how does it manage to move them in So many different directions? It's only pulling in one direction.

All the functions concerning the world's seas and winds are the spinning of the axis. And the different contours on the planet. Think All this space exploration is a waste of time and money. For money spent and being spent it would have been better spent on innovations on this planet. As this is the only way life has and will be a help to man. They keep moving the goalposts back. When they discover another planet, and they find nothing they say they we will find it on the next planet. This is to keep them in a job. Another way to waste taxpayer's money. Scientists has discovered that in all Europe there are only, six family lines. So European families travelled to the far corners of the world. I would like to know how many different family lines there are in the rest of the world. I was always bottom of my class in school, this was in all subjects. If I am wrong with my theory, so be it. Logically watch and see where the sun is.

The scientist are going to have the latest space ships spinning this will cause inside the ship to have its own gravity. Why does the moon not cause a gravity field? The reason that the winds are moving in different directions is because of the temperatures rising and meeting cold air most of the time; it is not visible to the naked eye. It becomes noticeable when the temperatures are in extremes then it causes whirlwinds, tornadoes and hurricanes. So, what causes the poles to change places over millions of years? This is because of the axes not being in an orderly movement. So, the tilting causes a jolting movement. Gravitation is caused by the planets spinning. That's why as Soon as you leave the Earth's atmosphere there is no gravitation, as you are in a vacuum, as with all the other planets. The reason that tides change is because of the Earth's tilt. This means when the Earth tilts the water stays in the same place. That causes the land to tilt this gives the effect that the tides are going in and out. Mountain and valleys Also help in the movement of the winds. The tilting of the axis is not on a precise direction as it has a twisting, jerking movement in other directions. So what causes the poles to change places over millions of years? See below Einstein's theory of relativity. As for time in space changing the age of people, so that apersons parents become younger than their children, this is not logical. My theory is that space time is not the same as human life span, this does not vary. This is what they have not considered.

If the winds on the Earth are usually up to 100mls per hour and the moon was controlling our seas, how does it manage to move them in So many different directions? Light comes down straight down, Sound comes down straight, So the moon's gravity must come straight down, if it's there. The physicists say that the sun's rays bend. If this was true then what stops them bending all the way around to the dark side of the Earth. The fact is that it is the particles in the Earth's atmosphere that reflect the rays. As the rays move from a straight line into a light that bends confuses me. All the functions concerning the world's seas and winds are the spinning of the axes and the different contours on the planet. I do hope that there are a few of my readers agree with this. As with most of my theory's/*logics* there straight to the point no going around the bushes

If the Earth is traveling at this terrific speed through space then why is the world not affected by this, there should be winds. This is Isaac Newton's law on equal force come in.

I was always bottom of my class in school, this was in all subjects. So if I am wrong with my theory, So be it. The scientist are going to have the latest space ships spinning this will cause the inside the ship to have its own gravity. Why does the moon not cause a gravity field?

Gravitation is caused by the planets spinning. That's why as Soon as you leave the Earth's atmosphere there is no gravitation, as you are in a vacuum as with all the other planets.

So what causes the poles to change places over millions of years?

EINSTEIN'S THEORY OF RELATIVITY.

As for time in space changing the age of people, So that apersons Parents become younger than their children, this is not logical. Where they have all got it wrong is that time is the same every where. It does not change. You might say that there are different times depending where you are on the Earth's surface

If you are in Paris and one hour passes, then one hour will pass here in England. The time doesn't change because you are in a different place. You might think this is logical, So why should there be any difference if you are in space. There isn't So I think that Einstein and all the other Boffins should look again at these mind-blowing logics.

If the winds on the Earth are usually up to 100 miles per hour, and if the moon was controlling our seas, how does it manage to move them in So many different directions? All the functions concerning the world's seas are the spinning of the axis. And the different contours on the land change the direction of the winds. They say inventors are mad. They don't know they're mad. I'm not stupid like them, I'm an inventor. I know I'm mad. I was always bottom of my class in school, this was in all subjects. So if I am wrong with my theory, So be it.

Logical As with most of my theory's/*logic* there straight to the point no going around the ice-bergs

Gravitation is caused by the planets spinning. That's why as Soon as you leave the Earth's atmosphere there is no gravitation, as you are in a vacuum, as with all the other planets. The reason that tides change is because of the Earth's tilt. This means when the Earth tilts the water stays in the same place. That causes the land to slide under one side of the Earth on the other side it travels above. Mountain and valleys Also help in the movement of the winds. The tilting of the axis is not on a precise direction as it has a twisting, gradual movement in other directions. What causes the poles to change places over millions of years? See below

EINSTEIN'S THEORY OF RELATIVITY.

As for time in space changing the age of people, so that apersons parents become younger than their children, this is not logical. My logic is that space time is no different than time on this planet. The human life span, this does not vary. That is what they have not considered.

If the moons gravity is controlling our seas, as the experts have been telling us. How is it that it can pull all the water in a sea in the moons direction and yet there should be other things that it can move? How is it for instance there is Something a lot lighter and should be more movable and what I am about to say Also applies to another body her on Earth. If the scientists and mathematicians can explain this to new followers of the universes I think they would be enlightened for this information. Can you see now the reason for this logical way at looking at things? There is a view put forward that the Earth's rotation has slowed down the moon's rotation, So that the same side of the moon is always facing us. What does this tell us, this is an important question? Thinking about what has just been said we can understand that the Earth's gravitational pull is stronger than the moons. This means that there is no way that the moons gravitational pull can affect anything on Earth. This Also applies to all the other gravity fields that the sun's and planets produce. I will explain later how I have come to this permutation. there is another fact that they say and that is the way are Earth's spinning on its axis keep the sea level at equal levels around our planet. This is a wrong conception of how the spinning of the axis of the Earth works. I hope you will be amazed by my working out of the way the axis spinning effects our planet. There is another spinning of the axis that geologists have not managed to see in all there studying of the Earth's layers and there contents, they have come to the conclusion that there was an ice age. It will be explained why this is not So.

How all the experts can come up with the same answer I do not know? Like in all trades and professional bodies the teachers, trainers, and any one that is there to teach any new personnel to their place of work, they teach them exactly what their predecessor taught them. The space industry is one that shows this more than any other trade or profession. They are taught and So what they are taught is passed down to the next generation. In all other businesses every So often Someone comes up with an innovation that is an

Bang Goes That Theory

improvement on how things used to be done. This is in all fields of industry. With space exploration this does not apply. As I have stated the new follow the old and nothing can change. There are more theories in what is going on in the universe then in any other industry. Maybe with my way of looking at things there will be a dramatic change in how the space exploration is conducted. I personally think that every penny that has been spent is a waste of money. They can carry on spending billions of pounds going to all of these far of planets and just what do they think they are looking for, aliens. And if they find them what do you think are Peace loving humans will do I think it's, lets destroy them they are a danger to us here on Earth.

If you have invented, produced or discovered Something new then that is the end of that. So why our Lord would put anything of less intelligence anywhere else in the universe I do not know. There are enough wonders on this planet as in plants, insect's animals and man.

I suppose you are asking yourself this is a lot of talk and no action. Everything will be explained when the time is ripe.

The physicists say that the sun's rays bend. If this was true, then what stops them from continuing to bend as it is still dark on the opposite of the Earth sun. My logic is that it is the particles in the Earth's atmosphere that reflect the rays to the darker side of the Earth. when the first sunlight is seen as the sun moves higher in the sky As it moves from the direct position less light gets through which makes it looks like a steady increase in light. All the functions concerning the world's seas and winds are the spinning of the axis's and the different contours on the planet. The Earth is spinning on an axis? What caused this, are we the only planet in the universe to have seasons? The lame reason I've heard it was hit by a meteorite. If this was the case there must be an awful large crater Somewhere along the Earth's surface. And if it was that large, it would have destroyed the Earth. We'll all other planets have been hit by meteorites and those nearer the Milky Way would have been hit by bigger asteroids. When I say bigger asteroids, they can only be a certain size and surly they would be classed as a planet.

A man on a railway platform sees a train in the distance going past at sixty miles per hour. It travels a mile in one minute. When he first sees the train he moves in the same direction. I could do a bit of arithmetic's and say he takes So long to move in that direction and how long it will take him. At this precise moment I do not think my mind can cope with working out any equations even if you find it easy to do by using your own figures. What I am pointing out is between the two speeds and the two distances although there is a vast difference. The man on the train has not got any younger than the man on the railway station. I hope this proves that this Also applies if they were doing these permutations in space.

The scientists are now saying that the planets, comets and asteroids are travelling at a faster rate and that the universe is getting larger and this is caused by the gravitational pull from sun's and planets being stronger.

RUBBISH

I wish to point out that the reason they are not moving away faster is because the galaxies that includes the planets, asteroids and the gases are moving at fifteen thousand miles per second. To enable them to move faster they would must have Something like the Big Bang affect. No that's wrong because if that was the case then it would mean that their speed wouldn't change it would have had to be a bigger sun. This is really complicating. Let me think? Got it.

This matter that is travelling in an outward direction cannot go any faster than they did when the Big Bang first started them on their journey. They are governed by this fact and there is nothing in the universe can make them change that. And they will, keep travelling in an outward direction being it in an orbital direction owing to the universes moving in an orbital spin. Let us remember when the first Big Bang accrued there was a massive amount of matter being blasted out. The gases, asteroids and planets that were created from this sun then created all the galaxies in the universe. When this sun exploded the astrologers say it created a black hole Also remember that the smaller matter was in their zillions and I should think a lot more. Now which direction do you think these asteroids would travel they of course would travel in all directions. They did not of course keep going in an outward direction if they did the universe would be an entirely different to what it is now. You might as well ask what the difference would be. The universe would of course be in a sphere not in an oval as it is today. This means that all gases, planets and moons would have these heading towards them and of course be bombarded by them. Think about it Some will be heading for the black hole that was created by their mother sun. As you will have noticed I do not believe in black holes, these comets will of course come out of these black holes. While they are in space they will take on ice. The physicists and scientists say that Some comets are made up of ice crystals. What is it that has come from the matter of a sun that can produce an ice comet? How have I come to this scenario? It is easy to understand with everything moving away at seventeen thousand miles per second then there is no way they can collide with anything else. So they must going in the opposite direction. That means that these black holes have no gravitational pull. Otherwise these comets would not go through. For anything to create a Penny Force, it must be a Solid as in a planet or a sun that is spinning. Gas planets might spin but this is because they are gas planets and are caught up in their sun's ability to make them spin. This does not give them the power to control anything around them.

Let's take the first Big Bang this is the Milky Way as the sun's from this occurrence have left the gravitational pull of the Milky Way. So as all the planets, asteroids and comets are Also moving away. If this is the case then there is going to be a time when they all reach the outer reaches of our universe. This cannot happen. Do they all stay on the fringes or do they manage to escape as all of the previous happenings have moved away. Remembering that they will all be smaller sun's and planets, they of course will be travelling at the same speed. What do you think would happen if Some galaxies travelled faster than others? What would happen is that the galaxies would be more spaced out in the universe. That would make it easier for our telescopes to see what was going on. There would not be a cluster of galaxies in the Milky Way as there is today.

Evolves they are all was talking about species dying out wolves bears Evolve in decades not in millions of years I have just heard (feb/march 2015) that they are saying that light is travelling at a faster pace then what was originally thought. I went outside and waited for the sun?? When it rose, I saw it right away I don't know what speed it was travelling at but it was quite fast enough for me. In fact, I saw the light before the sun rose does this mean that my eyesight is faster than the sun's rays.

Please note the water on the other side of the Earth does not go down the sink in the opposite direction to England's. As you have moved to the opposite side of the centre of the worlds spin.

As the centre of spin, applies from the centre of the Earth not the Earth's surface.

Please note the water does not go down the sink in the opposite direction to

Craters all planets have craters the reason there are no more craters hitting the Earth are because the objects are fore ever getting smaller/ Also the spinning causes the objects to enter at a less direct direction. I don't possess a telescope So I must rely on television coverage. I have noticed that the collisions made by comets or asteroids hitting the moon that there are no indications the they have struck the moon in declining direction So where do the scientist/physicists get the idea that a big comet/asteroid struck the Earth to give it its axis spin as the Earth and the moon where created by the Big Bang. The moon and Earth were created by God. That is why they belong to the outer edges of the Milky Way.

The scientist/physicists say that Neptune has the strongest winds in the universe. They say they are 1,500 miles an hour. Does this mean that they have a lot more moons? They say moons gravity pulls the Earth's tides that cause them to rise and fall. If there was water on Some of these planets and they have a lot of moons as in the case of Jupiter which has 67 known moons. Can you just imagine the scene? The family have just arrived at the beach and the mother calls out to the children, keep your eye on the tides there will be part of the sea where the tides going out and further along the tide is coming in and then out. Now you can see my problem of our moon affecting our tides.

The milky way/universe is spinning if go by what the scientist/physicists have come up with the Earth's spin is due to a large meteorite hitting the Earth and putting it in a spin. That's got my mind in a spin. Let's look at it in connection with snooker balls. The Blue ball Earth is hit by comet ball red I direct hit this causes a crater. Why are all the craters appear to be a direct hit and none appeared to strike the Earth a glancing blow causing a skidding effect? You can see this by the moons The Earth is travelling at 17.000 per second through space. The object on impact this would Also

Bang Goes That Theory

Let's look at it from the beginning. There was the sun. (More about this later.) Do not look directly as you would be in trouble. The physicists/scientists say that the sun got hotter and hotter and it got So hot it blew up. This is not how things work. The original sun was in space a (vacuum) it was of molten plasma containing all the metals that we have on this Earth. Let's look at Something we can compeer it with. A pressure cooker is a fine example. Say there was no pressure outlet, what would happen? You've got it would explode. Now say you had this pressure cooker in a freezer. I am getting a head of myself. If you try to defrost a freezer using hot water what happens it that a film of ice appears on the inside of the freezer. This cold temperature in time will form a crust then holds the heat in. So it gets hotter, then it gets So hot it explodes.

The following information was found on secrets of the bible. Raiders of the lost past. Series 3 epiSode 1. The reporter was an astrologer called Tudor Parffit.

There is a tribe in Zimbabwe South Africa called the Lemba tribe they claim that they travelled from Egypt thousands of years ago. This was during the time when the Jews had to leave because of persecution. They Also brought with them the tablet that God gave too Moses. They carried it in drum as this was not in the original cask as that was a metal cask. The bible states that it was transported in a drum. When you think about it the original ark would have been too heavy to carry remember they were fleeing So they would only take the important things. The tablet was transported in a drum. This is mentioned by the elders of the tribe is known as the Lemba tribe. The leader Matriva is one of seven priests in this village that their D.N.A matched the Jewish priests in Israel.

There are tribes in Kenya and Uganda the Turkana tribe that rely on camels as part of their life style. As we know camels come from the east So they must have brought them on their journey. Their exodus from the Egyptians they could bring anything with them that they required.

The physicists/science say there must have been another race that can build the pyramids in the mountains of Mexico. There were numerous cities built by different tribes.

There is a copy of this drum is in the Zimbabwe museum.

The DNA was taken from the priest of the Lemba tribe. Fifty per cent of the DNA was found to be the same as priests/rabbi's in Jewish population.

How did that come about that?

The following information was found on secrets of the bible. Raiders of the lost past. Series 3 episode 1. The reporter was an astrologer called Tudor Parffit.

There is a tribe in Zimbabwe South Africa called the Lemba tribe they claim that they travelled from Egypt thousands of years ago. This was during the time when the Jews had to leave because of persecution. They Also brought with them the tablet that God give to Moses. They carried it in a receptacle that was more in keeping with time. It was carried in a drum. The original ark was a metal cask. You can understand why this was left behind. It was heavy, and they were not going to spend any time hanging about.

This was in the form of a drum. The bible states that it was transported in a drum. When you think about it the original Ark would have been too he must carry remember they were fleeing So they would only take the important things. The tablet was transported in a drum. This is mentioned by the tribe known as the Lemba tribe. The leader Matriva is one of seven priests and fifty per cent of their D.N.A matched that of the priests in Israel. There is a copy of this drum in the Zimbabwe museum

The physicists/science say there must have been another race that can build the pyramids in the mountains of Mexico. There were numerous cities built by different tribes.

And the bible says that the meek shall inherit the Earth.

Trust it to be now when the politicians have made a right mess of everything.

Modified crops. Are they needed? In the beginning, this applies to the Garden of Eden. There were no pests, bees, or any other form of flying insects; this was before Adam was enticed to eat of the forbidden fruit.

Pesticides verses modified crops.

You won't believe the advantages of one against the other. I hope this information will change all those that have the view that modified crops do harm to Some expects of the environment. Yes it's to do with the environment and why the modified crops are better for the environment.

When you think about it the farmers spray the crops with insecticide to kill the pests, all this pesticide all finish up in the Soil and then they finish up in our rivers, and then down to are waterways. Then finally they finish up into the seas poisoning everything they come intact with. There is none of this with modified crops. Now let's take what the environmentalists will say. Their argument will be that by having modified crops there would be no insects for the birds to live on, and what about the bees? Well the farmers spray the crops and So do the gardeners. This has not affected the amount of insects. And there are millions of insects and I think they all came to my garden. I forgot to spray my fruit trees, fruit bushes and my strawberry plants. Does this mean I am now working to help the environment and keeping the planet in a healthier place? I thought I would sit back and let nature take its course. I mean that's a dropped clanger. Let's take the green fly. When the green fly lays its plasma say about two. This is on each plant it visits. And it is not over a period it goes straight away to the next plant. In my case it goes to the next strawberries plant. These two aphids have the capability to have young plasma each day. That's six, the next day it's eighteen. In a weeks' time there is a hell of a lot of green-fly. Now knowing how nature works it is providing a predator, the lady-birds. They are not around on the first day as this would mean after one good meal there would be nothing left. Kaput, we now have no greenfly. The lady birds look around for their next meal, nothing what we have is a lot of dead lady-birds. As there can be six-ten stalks with flowers on this is a hell of a lot of greenfly. The lady-bird has its hands full, Sorry its mouth full. Now if this lady-bird is going to do its job properly and eat the entire green fly on the strawberries I think it's going to have a job taking off. I don't know how many greenflies a lady-bird can eat but I think it's going to lose the battle. I hope you now get the point; nature does not do the job properly. These insects as are weeds were our punishment of Adam eating the bloody apple. (Thank god Eve got Adam to eat the apple. there is more about this later.) If the number of insects that were required to do the job properly you can forget about the plagues of Egypt that would look like an annoying fly buzzing around, you in a severe winter. The air would have So many you would be swallowing them by the dozen every time you took a breath. If the farmers and the gardeners didn't spray the crops were would we be. I don't hear the environmentalists complaining about all the pesticides that are being used. There next argument could be the modified crops would affect other crops and contaminate them. If the scenario above about the ladybirds having no more greenfly on the menu and that was the end of these two insects and it happened with all other predators and their food supply would this be a bad thing? Let's take you back a long time ago. I mean a really long time ago. In the Garden of Eden there were no insects and there were no weeds. Which means that the insects were not needed to pollinate the crops? So what was there? There is Something that gets everywhere; and it is more efficient than anything else.

MODIFIED CROPS.

Are they needed? In the beginning, the following applies to the Garden of Eden. There were no pests, bees, or any other form of flying insects; this was before Adam was enticed to eat of the forbidden fruit. That was when man was cursed with the above. We then have Adam and Eve. They must have been the first to start agricultural farming. The reason we know this is because when Cain and Abel took gifts to our LORD he was pleased with Abele's gift of a sacrificed lamb. He was displeased with Cain's offering of cereal. This was the reason for Cain killing Abel. We now have these insects in every climate all over the world. Leaving this to the side for a time let's look at other benefits. With these poisons no longer in the environment there would be a lot less people attending the hospitals. This would be a Godsend. Just think of the amount of insecticides that are being pumped into the air and ground. We would certainly have a healthier population. Can anyone explain to me why there are millions of wild plants, and there are no insects that live on them to the extent that all are destroy? Not even the green-fly attacks them. When there is an outbreak of this kind it does not destroy them all why is this. How about slugs, they really have an appetite for my lettuce. The female slug lays about fifty eggs. The same thing happens as it did with the green-fly. The birds would have a job catching them in the undergrowth. There are no ponds in my neck of the woods, So no hedgehogs. When did these slugs evolve that only like my vegetables and fruit? The gardener puts down slug pellets. The rain and damp cause them to dissolve into the ground. That's more chemicals/poisons into our environment. (If things I am working on go my way it will reduce most of this poiSon that's entering our environment.) What are the down sides? I can't think of any. Sorry I have just thought of one honey. Sorry do not take that as a friendly greeting. It's the honey that bees produce. I am a regular user of honey I use it every day in one way or another. But if it's going to make our planet a healthier place to live in then So be it. I am willing to give up this luxury. The crops would be able to grow to their full advantage. If the birds and other forms of wild life as in ladybirds did their job properly we would not need insecticides. If they were anything near being good at it then their environment would be like the animal life as their food supply got less their young would not survive So their prey would increase, and their predictors got less and So the cycle continues its only man that can destroy the animal population. You're probably thinking I have forgotten all about the magic remedy. What can it be that will do what insect and birds are doing (their not very efficient) What is this magical thing, there is no machinery no insecticides involved, No workforce, the advantage to the atmosphere must be tremendous. Okay we have modified crops and no insects. I suppose this will affect the producers of insectaries and fly sprays. I will now tell you what will take over from the insecticides. It is not Some magical potion and it has been with us since the beginning of time. It's the winds. Have you seen the pollen in the air on a hot breezy summer's day then you have got the point? This applies whenever a plant is starting to produce pollen. One gust of wind and thousands of pollens gets distributed over the whole world. This is by the way much more efficient than any insect or bee can do. As far as what we will miss if there were no birds. I can always whistle if anyone is getting depressed. Now let's take the bee. By my deductions we don't need these. Okay I like honey I use it every day. But I suppose it's a small thing to lose when you think about the advantages. We control the bee's home life So what's wrong with having areas that are controlled for them to obtain pollen. I suppose in time we be able to manufacture honey. There are lots of more substitutes like maple syrup. So, bring on modified crops.

Dinosaurs

We have lost the dinosaurs and much later the dodo. Has there been any disadvantages to man I don't think So in fact with the way man has been destroying the animal's environment and they are now in zoos or invading our habitat. If we did, we would certainly need Some very big whoopee bags. What effect has this on us, I am glad the dinosaurs have gone I have enough problems worrying about bee stings. And just think how long it would take to barbeque a dinosaur leg. Still an egg should satisfy the whole family. And as far as the slow-moving dim-witted dodo is concerned I suppose women could always get their husband to take his place. Let's look at the dinosaurs. The archaeologists say that there was sufficient food for the dinosaurs, one reason they say was that the Earth was hit by a meteorite and they all suffocated. If this was the case, there would have been signs of ash all over the Earth at that time. There is nothing else that they can come up with. Don't think I am being presumptuous but how about a flood. I mean to they say we had an ice age. (No, we didn't. I will explain this later.) For every negative there is a positive. So, the ice melted, forbid me but how about a God. Up go all the arms as though in disbelief that Somebody thinks there is a God. There are far too many things that the scientist and physicists can't explain, and most of them admit there must be Something else. It would not be the only time that God caused a flood as the bible points out. This has happened a few times until he decided to leave man to his own devices. Can anyone tell me that in the beginning there was nothing how can you create anything from nothing.

SPEED OF LIGHT

The physicists have come forward with the scenario that when you travel at the speed of light to another planet and you come back you will be younger than your children. Wrong In fact in one scenario if it takes you fifty years to go to a faraway planet and fifty years to return when you get back to Earth, millions of years will have passed. Wrong. I am no intellectual, but this is not possible. They try to show this in different ways and if you just take this at its face value it seems right, So I decided to start at the beginning. As life has gone by I realised that when you have a problem you go back to basics. This applies to any problem whether it is a sports person who as lost his ability to play to his usual standard or a

Bang Goes That Theory

maths problem. I had a friend once, don't get me wrong I have more than one friend I think at my last count I had two. He said he was always getting mixed up and when he came to putting in a b or a d he didn't know which one to use. I told him just go to the spelling of a dog or a ball and you will then know which one to use. I believe any problem can be Solved if you go the right way about it.

Most of what physicists' try to explain is above my head. I then try to Solve the easiest one. There is one they demonstrate concerning the speed of light. You are contained in a box on a train. The train is travelling at forty miles per hour. In the box there are two mirrors, a light passes from one mirror to the other and back again. This is the speed of the light. the speed of the train is travelling at forty miles per hour, So does this mean the light travelling between the mirrors is travelling forty miles per hour faster than light. They are looking at this in the wrong light. Pun intendered, these are two different equations. They must be separated; one has nothing to do with the other. Let's roll a ball across the train's width. This is a wide train. The forward movement of the train has nothing to do with the sideward movement of the ball. If it had, then the ball would Also move in the direction of the train. If you look at it from the physicist's calculation, then the ball would be travelling faster than the train. What a clever ball. This has not been repeated it's a different ball.

Another scenario the physicists give is a man standing on a train station he sees a man on the train and he is firing of the gun. I thought if I can Solve this one then it will apply to the speed of light.

The physicists set out a mathematic problem, to prove that light is not at a constant speed. They use the following scenario. You are on a train that is travelling at sixty miles an hour you have a box you have a light that is bounced between two mirrors and then it bounces back to the first mirror this light is travelling at the speed of light. Now these mirrors are on a train that I am travelling at sixty miles per hour So the light is travelling faster than the speed of light. Wrong.

My theory/ logic Something that is on a form of expulsion does not relate to the forces outside the train. In Einstein's relativity that light is constant no matter whether it is being observed regardless of the observer.

Another way that they demonstrate this is a train is travelling at forty miles per hour a person on the train fires a revolver. Say the bullet leaves the gun at sixty miles per hour. The bullet must be travelling at one hundred miles per hour. Wrong. If the person fired the bullet in the opposite direction then using the said account it should be going the opposite way.

Our sun has ninety five mass of all the other matter in our galaxy.

How come the moon shows no sign of volcanoes only asteroids/craters?
The milky way formed by the first Big Bang. This was the sun that was created by God. There are no other circumstances that the physicists or scientists can come up with. No matter can multiply from nothing or from itself. This Also applies to life. So all the origin planets asteroids, and meteors, came from this Big Bang. That means that they where all smaller. What caused the sun's to explode? The physicists say they grew hotter and hotter to such an extent that is what coursed it. Let's take a look at space before the sun entered into the scenario. The temperature must have been at its coldest (need to find the coldest) then mysterious a sun appeared. This then produced gases, and dust particles. Let's take a look at what happens when you have heat and cold meeting. We know the sun has been with us for billions of years and it has not heated our universe let alone space to any degree. Like all micro waves they can not be seen. The sun's micro waves can be seen when they come in touch with anything, Such as planets, asteroids, and gases. Let's get back to the sun and the cold in space and what happens.

If it is proven that my logic is wrong then it is a theory. That should cover any clangers I have made.)

If the original sun was created from gases where did they come from they have things the wrong way around. You must have heat to create gases. This sun got hotter and hotter. The physicists tell us and that was the cause of the Big Bang. Our sun is cooling down they can't have it both ways. When the Big Bang occurred how gravity did continue to keep the planets together. So what is causing the universe to continue to spin? My theory is God created the original sun. So now we have the original sun that is in a vast space. Space must have been at its coldest. As there was nothing to warm it up. This caused the sun to cool down.

GRAVITY

It takes 2g's speed to leave the Earth's pull of gravity.

The scientists are saying that the universe is cooling down and one day there will be nothing left only cinders.

This makes my logic of the universe correct. The origin bang did not occur because the sun got hotter and then exploded. It got colder before our lord created this sun the space was at absolute zero. The moment the sun appeared there was a distance from the sun that staid at absolute zero as the sun was not getting hotter then it must have been cooling, and So continued. In time the crust would get thicker and thicker. In time the sun would start getting a crust on the surface. This is how the Earth is now. As Earth has past from this stage to the next stage thing to happen would appear volcano's and Earth quakes. That is what would happen to all the sun's. The Earth is like a pressure cooker the volcanoes and earthquakes release this pressure. So if the global warming is not stopped then Earth will explode. The bible says that fire and brimstone will rain down on the Earth. So this global warming will not be stopped.

How the first black hole appeared??

I had to put my thinking cap on where black holes have come from. I have given my logical view concerning their expansion. So where did they come from. You must go back to the beginning. I learned this from watching and listening to sports coaches. Being a snooker fan when a professional player is not playing up to his best abilities he advises them to go back to basics. That takes us back to the original Sun.

I have explained my logic for the reason the Big Bang happened this was by the cooling down of the sun. So let's take a closer look and think about it and what would happen. This would happen in a nanu second, or quicker, if you go by what the scientists and physicists say that there is no limit to how small or large that any thing on this planet can go to. If this is So why is there a limit on the speed of light?

This Sun's rays, and the gravity that it had created extended into outer space. Let us think about this for a second. We have a Sun, as I have said before you can't get Something from nothing. That only leaves one thing. There was a divine force that must be a God. Bear with me for a minute. Let us put ourselves in that position. There is only space and that means that there is no light. You must think that the logical thing to do would be to create a form of light. As you can see from the start of this paragraph we have a sun and nothing else. Then it exploded. All the components, gas plasma, and Solids would burst out into space. This would leave a void. More about this void later. We can see what is happening and if this carries there would be no change to this scenario. That means we must stop this happening. What we need is Something to stop this happening. So we cause the Sun to spin, this causes the matter to slow its outward progress. The scientist and physicists call this gravity. They Also say that the gravity that is produced by a black hole draws all matter into its field and nothing can escape, not even the sun's rays. Come on lets be sensible, who in there right mind believes this. And I don't mean any one that is getting paid to study space and all its wonders. What is it that makes me So sure, well I am sure that all the people that will see that my explanation is So obvious that there is no need to explain it any further. The sun's created by this bang travelled at 17.000,00 miles per second. As did all the gases, asteroids, the sun's travelled away from the original sun's position the sun's rays and gravitational pull began to pull the planets, gases and asteroids and all other material So creating Solar system. That would mean that the rays would be travelling through there Solar system. As the rays reached the next galaxy, it does not mean that the brightness would be twice as bright, as each ray would counter balance each other. This Also applies to the force field (gravity.) now lets look at meteors. I do not know much about meteors. As I have just said sunlight and force field counteracting against each other why does this not apply to meteors. Meteors are the scientific explanation is that they come from the asteroid belt between mars and Jupiter.

At the time that the Big Bang happened, all the Sun's components where blasted out into space. As in a gas form Also winds, plasma, planets, asteroids and of cause the largest object would be sun's. The sun's being the largest and having the same abilities as the original sun, they would have a magnetic field, and what the scientists call gravitational as I have said previously gravity does not draw objects closer to the planets surface, it holds then in their orbit. I call it a force field. The outer part of the sun would of cause be the Solid rocks. The inner part would of cause be the molten plasma. They are all moving away from the explosion at seventeen thousand miles per second. As they all travelled outwards there were far too many for me to put a number to. I would say that there was a lot. The Solids would be the first to leave the explosion area followed almost instantly by molten *plasma*. Now all these sun's would have the same forces as the original Sun, as in a sunlight force field and magnetic fields if this is So where the sunlight magnetic force and force field met they would cancel each other's power out So they all would stay with their original power. This is Isaac Newton's theory that two forces of equal force will rule out each other. To get back to the question of all matter travelling away from each other, and as what the above says. The sun's would catch the planets and asteroids in its *force fields,* and they would orbit around that sun. As these planets and asteroids where on the same plain you would not get a three D image. And that is why all galaxies systems move in a spiral movement.

God as everybody knows is not of human form. It is beyond mans imagination to even try to understand whether he has been around since before the universe or there was Some magical occurrence that happened.

Einstein said that he wanted to find out the basic of this form that could travel at the speed of light. And maybe our God would be the only thing that could travel faster than light.

And God, 'said let there be chameleons,' and there was, after six attempts he give up.

And God said to Job no one must look back on their past life.
As they where leaving Sodom/Gomorrah, Jobs wife turned to have a last glimpse of her home. God was displeased and turned her into a pillar of salt. All that's needed now is the vinegar and the fish and chips.

This scenario is Also correct today. You only improve life by looking ahead. You cannot change the past what you do today is your future, So make sure that what you do today you will not regret your future, tomorrow, for when tomorrow comes it will be to late to do anything about it.

BLACK HOLES.

If the gravitational is So strong that sunlight cannot escape, yet they say gravitation is one of the weaker elements. For every positive there is a negative. That means sunlight is a negative until there is a positive that can show there is sunlight. Any matter in the sky such as gases would do this. This Also applies to gravity. If we go by the first sun's gravity and sunlight was travelling into space billions of years before the Big Bang and that is why as the universe expands you can see the effect of the sun and gravity. Sunlight loses its heat and So the sunlight is not as strong but gravity gets stronger let us look at a simple way that we can understand how this is possible. Take a simple thing like a wheel. If you spin it the circumference of the wheel travels faster than the part nearest the hub. So planets that are furthest from the sun will travel faster. Now Also if you think about this then there is less likely that matter can be drawn closer to its surface.

The astrologists, physicists and the scientists explain this.

Does a torch light have radiation I think not So how does it light up colours if the colours are the effect of the sun's rays? Not logic. It's not the different colours of the sun's rays that produce colours it colours in the elements that the white light shows their colours.

THE MOON

The moons creation was created at the same time as the Earth. And if we go by what the bible says it was god's creation. They say that a planet the half size of mars hit the Earth and a big chunk got dislodged. You have had my opinion on why this is not possible. Let us look at the dynamics of this scenario? When an object hits or crashes into Something we must take into consideration the structure of the object that is being hit and the structure of the object hitting the said stationary object, Earth. Things hitting planets cause's craters these are different to craters caused by volcanoes. These have a considerate higher peak than what an object hitting it would cause. This can be seen in the differences between Earth and the moons contour. There is another distinction that can easily be seen in the difference between the two. A volcanic eruption shows plasma flow that flows over the side this leaves a river type impression on the side of a volcano.

I cannot see any form of the craters on the moon resembling volcanoes. The moon is covered by objects hitting it, while the Earth is covered by volcanic eruptions. We know that meteorites are always hitting the Earth

If the moon was originally part of the Earth how come their surfaces are So different? Taking logic at its face value they should be near enough the same.

from the photos. an object hitting the Earth it said to have been one half the size of the Earth's or one quarter by different bodies in the building of the universe. Would any of my readers like to give a size of this object? It does not matter if you are nowhere near the right size as you will be doing as the physitrists do guessing I am Sorry you cannot be paid for this guess as you are not classed as a physiatrist. The boffins say that this occurred in the early years of the Earth's appearance. We will now go back to the early years what do we know about this. We know or have been told that there was no ozone layer and as the photo indicates it was a direct hit. Can anyone of my readers see this impact knocking a chunk of the Earth? I will use Newton's law on equal forces. This is what I think would happen. The offending object would penetrate So far into the Earth's surface removing the crust giving the same appearance as a large comet or asteroid would make. Let us pause and then rewind. This was supposed to have happened as was stated before. There is another hole to this scenario. Pun in tendered. When the universe was first formed it was from a ball of molten plasma and rocks. This is if you take it from the astrologer's point of view. Not a hole to be seen. That has blown a hole in the moon being formed in this way. There is another good reason why this goes against all the law of physics. Why the astrologers. Physitrists and scientists have not seen this I do not know. Oh my goodness I am getting good at these physics.

Have one got an idea what that is. The phoenix probe sent back pictures of Mars surface. It showed an asteroid that hit its surface billions of years ago. This asteroid was So small that it could be hand held. If there are asteroids that are the size of Mars shooting around the universe how come our techknowledgy have not picked them up.

They say that a large asteroid hitting the Earth will destroy it. If an asteroid the size of the one that was the Source of our moon did not destroy the Earth what size do the experts think this will be. How is this for anther revelation? The universe was in place as it has been since the Big Bang. We all agree to this. When I say we I am including the physiatrist, scientists and the astrologers. We will take this to our nearest occurrence. That is our Solar system. We have our planets which consist of mercury Venus and Mars. Then we have the two gas planets Jupiter and Saturn. Lastly there are the two ice planets Uranus and Neptune. As you will have taken notice that the heaviest planets are nearest the sun. That means that in-between Earth and Mars there is nought. Then this chunk comes from nowhere and gives the Earth a clout. A chunk of Earth gets flung into space. The largest size this can be is the same size as the object hitting it. Now we do know the size of the moon and that it is 800 miles at the diameter, and Earth's is 2,160 miles at its diameter. This is one quarter the size of the Earth's. Mars is 400 miles this is half the size of the Earth. I will try to work this out So that you can understand what I am trying to do. I will need my co writers to confirm this total.

This means that going by the size of the moon and the abilities of our Solar system then the moon should be the other side of Mars to the Earth.

Was our galaxy formed after all the other galaxies? So by the rule of mathematics'/physics the Earth and moon should be smaller in compariSon to all other planets. They should be smaller.

Radiation, nuclear plants they say you must contain radiation otherwise it continues to expand. Why when the atom bombs were dropped did it not keep on expanding, Also as in Chernobyl?

SOLAR SYSTEM

The majority of Solar system is spinning in a clock wise direction; this is as seen from the direction of the Earth. Each galaxy has different gravitational force. This determines the size of their circumferences as in the space they control. Where the gravitation force of three or more galaxies meets it creates a void. As the spinning of this Solar system two things happen. Any material in each galaxy are kept in that galaxy, by two factors one the gravitational force of that planets gravitation and the force of the other two galaxy system pushing the material away. The material that would be in this void would most likely be gases and maybe Some small particles of dust.

A black hole does not mean that the black hole has such a strong gravitational that any light or any other material that enters it cannot escape. The reason for this will be explained later. I am no expert but in the universe there are lots of places that this can happen, not just the space between galaxy systems. Let's have a look as this from when the Big Bang accrued. The original sun exploded. This was caused by it getting hotter.

MILKY WAY

All sun's and planets consisted of molten plasma.
And all of these are supposed to have a Solid centre as predicted by scientists and astrologers I think that the scientists have put their two penny worth in. My logic/theory is, that the centre of them, burn up all the oxygen and the iron cools down until it is a Solid, therefore no oxygen no heat. That means they lose their ability to burn So its logic that they cool down. This is using physics as an explanation. This does not commute the iron or other metals would come together but not in one block. There would be clumps at different places in the plasma and that is why it is found here on earth at different places.

Now we come to the crunch. D.N.A. A profesSor travelled all over Europe, Africa and Asia. The object of this was to find if there was any truth in the story that an African tribe in the past were the tribe that rescued the arc when they were allowed to leave Egypt. This was the exodus from Egypt. In all the countries by testing for D.N.A, we have come to the conclusion that we are all related. If the test the tribes and original inhabitants they will find that their D.N.A, matches. We know that the tribe in Africa is black but their DNA matched the priests in Israeli this would not be the first time our Lord had done this, it happened when man on earth was being treated like Gods and building huge monuments to the sky. This was to confuse the different tribes So that they could not understand each other. Your being ridicules I hear you say again. Have you forgotten Something? You must believe in the modern scientific discoveries it's the D.N.A. this cannot be wrong. They have done D.N.A. tests on the native Americans and have found that their D.N.A matches the D.N.A. of people in Europe. Americans in South America share ancestry with native peoples in Americans in South America share ancestry with native peoples in Australia and Melanesia.

But the two groups came to different conclusions when it came to how that DNA with ties to Oceania made its way into the Native American genome.

AUSTRALIA AND MELANESIA

But the two groups came to different conclusions when it came to how that DNA with ties to Oceania made its way into the Native American genome. The above paragraph states that they can not agree how the movement from Americans in South America share ancestry with native peoples in Australia and Melanesia. The two groups came to different conclusions when it came to how that DNA with ties to Oceania made its way into the Native American genome. The point is that the DNA is another proof that what the bible says about our Lord dividing the Earth's mass to separate the human race is another fact.

Gravity waves from two black holes colliding. By my logic of black holes this is not possible.

The waves that they have picked up on the billions of pounds spent on this project as with all the other billions spent on space expiration. If all this money was not wasted on these procedures just think of all the good it would have done can any one explain to me on what course is this taking us. And just what benefit has it done for one perSon on planet Earth. Anything that the scientists say will not convince me. Einstein didn't spend billions of pounds and a lot of his theories have proved to be right. (It's a pity I have blown them into the blue yonder.) Getting back to my observation of black holes,

BLACK HOLES

Black holes eating up the universe if any one knows any thing about the universe they know it is expanding. As this is the case and one thing we can be sure of. How do the space Boffins come up with this scenario.

They have worked this out because they see what they want to see. And that is the blackness is increasing. This is not So the other objects are moving away the space that is being seen has no matter So it shows up as a Black Hole would be seen. This is what they keep saying So why have they decided to change this format is it just to confuse us mere mortals. They would not like it if we had a mind of our own.

Concerning the different points of view about where they came from. The scientists and physicists have come to the opinion that black holes come from.

WHAT IS A BLACK HOLE?

Chances are, you've come across the idea of a black hole in science fiction books and movies, but these objects exist in real life too, and are just as fascinating. Black holes are objects in the universe So massive and with such strong gravity that nothing can escape them once trapped in their gravitational fields. In fact, most black holes contain many times the mass of our Sun. But even though they are quite massive, an actual singularity that forms the core of the black hole has never been seen or imaged. We do know they are there by the effects they have on the matter surrounding them. And we are able to see the material that they suck in through their tremendous gravitational pull.

Let us bisect the above and see if I can put it in a format that the average perSon can understand. The idea that the universe contains black holes came from. Eisenstein's theory and the comic books jumped on the band wagon and then the film industry took over with there outer space films. Years ago we got the idea that there was aliens coming to Earth and taking over our planet. We now know that this is a fallacy. You still have Societies that are saying Earth was populated by aliens and the gullible donate money for this cause. Sorry I digress. As the above states that black holes are a fact. But we can't see them, that is a big help in understanding them. Don't get me wrong there are quite a few things in the universe that we can't see but we know that they are there by their effect on other matter. Gravity is but one, the sun's rays is another, but we can see the effect they have here on Earth. Black holes now that's another thing we can not see them and we cannot see what they are pulling into their centre. If this is a true fact then every thing is going to be pulled into a black hole. Then we have the opposite this is a known fact and that is that the universe is expanding. Which one are we going to believe? I will go for the second scenario. You might ask why. Well if the first scenario is right and it can not be and why do I think this. The NASA space probe of the first photographs taken, show that the universe is expanding. The reason photos show that the matter in space has travelled outward. Do you need any other explanation?

THE STRUCTURE OF A BLACK HOLE

The basic "building block" of the black hole is that singularity. It's a pinpoint region of space that contains all the mass of the black hole. Then, there is the region of space surrounding the black hole from where light can not escape, giving the "black hole" its name. The "edge" of this region is called the "event horizon". This is the invisible boundary where the pull of the gravitational field is equal to the speed of light. It's where gravity and light speed are balanced.

The event horizon's position depends on the gravitational pull of the black hole. You can calculate the location of an event horizon around a black hole using the equation $R_s = 2GM/c^2$. R is the radius of the singularity. G is the force of gravity, M is the mass, c is the speed of light.

We will take the last paragraphs as true. As I worked out that gravity travels at the same speed as light, but it has a longer life. The physicists have not got to the point were they can work this out. Then again they would change my deductions into a theory.

Gravitational collapse is the contraction of an astronomical object due to the influence of its own gravity, which tends to draw matter inward toward the centre of mass. Gravitational collapse is a fundamental mechanism for structure formation in the universe. Over time an initial relatively smooth distribution of matter will collapse to form pockets of higher density, typically creating a hierarchy of condensed structures such as clusters of galaxies, stellar groups, stars and planets.

A star is born through the gradual gravitational collapse of a cloud of interstellar matter. The compression caused by the collapse raises the temperature until thermonuclear fusion occurs at the centre of the star, at which point the collapse gradually comes to a halt as the outward thermal pressure balances the gravitational forces. The star then exists in a state of dynamic equilibrium. Once all its energy Sources are exhausted, a star will again collapse until it reaches a new equilibrium state.

The above statement about Black Holes is a load of rubbish.

A BLACK HOLE IS BORN

Here I go again, using a few of my grey cells up. As I might have said before So not wishing to leave anything out I will put forward my observations. Gravity does not draw things into its centre; gravity holds objects in their Solar system or their galaxy. There is another thing that I or any other body that deals with space exploration has seen what happens when two or more galaxies meet in the universe. So I will put that right. Every galaxy or Solar system has a different strength. That makes them have different size circumferences. The larger ones will of cause be the stronger ones. Where two or more galaxies meet there boundaries will adjust when their strength are of equal strength, as in Isaac Newton's equal forces balance each other out. There is Some thing that happens when this comes about that should astound the physicists,' scientists and the boffins there is a void and this void contains Some thing that the bangs caused and these voids will not be in any uniform design to anything in the universe. It is of cause the gases. The Boffins say that the centre of the Milky Way is that hot because of the gases in there. As you have probably learned from my deductions gases are thrown away from sun's gravity and not drawn inwards. The hottest part of the Milky Way is not gases as gases would have cooled down until it is imposed by another point I might have said before concerns the physicists and scientists theory that a sun gets that hot that it collapses into it's self. It does not all the sun's will react the same way. And this is based on going by what all the experts have said. My gosh you might say what have the experts missed. They and I think the average perSon will agree that the Big Bang was the first sun to disintegrate. And how did the end come to the hottest and biggest sun that has ever existed did it collapse or did it explode. It exploded I would take that as a sign that the theory about a star collapsing is just a theory.

I would like to put forward a theory concerning the Big Bang and the milk way. We have all agreed that the original sun exploded. And as I have pointed out this was because of the cooling effect from the absolutely zero temperature. The crust must have got really thick So when the heat and pressure got that great that the crust went hurling into space. Now what we have left is a boiling cauldron of plasma. Now comes in Isaac Newton's theory of equal forces. You would then get this force being applied from all sides of the circumference. So maybe that most of this plasma stayed where it was originally and only Some of it went into space. This then would show as the hottest part of the Milky Way. As we look at the Milky Way we see that it is not in a complete sphere. This could be explained that the sun's crust was at different thicknesses and would extend to different distances from an explosion. I don't know if this makes any sense but it's only a theory. It goes against all my logic that when a sun explodes it leaves a black hole. Maybe the two are right as the first sun was really big. At least I hope that my theories do have a bit of logic to them I hope.

STAR FORMATION

An interstellar cloud of gas will remain in hydrostatic equilibrium as long as the kinetic energy of the gas pressure is in balance with the potential energy of the internal gravitational force. Mathematically this is expressed using the virial theorem, which states that, to maintain equilibrium, the gravitational potential energy must equal twice the internal thermal energy. If a pocket of gas is massive enough that the gas pressure is insufficient to support it, the cloud will undergo gravitational collapse. The mass above which a cloud will undergo such collapse is called the Jeans mass. This mass depends on the temperature and density of the cloud, but is typically thousands to tens of thousands of Solar masses.

MY THOUGHTS, MORE RUBBISH.

In the above statement about stare formation I do not have a clue what they are talking about. All I can go by is what happens to gases here on Earth. If any material is put under pressure than it's weight cannot increase. What happens to it is it expands outwards. This would occur more So with a gas. Gases do not explode into a Solid. The gases around the sun are put under the hottest temperature than any where else, and yet it does not explode. Maybe the absolutely zero temperature cools them down to the extent that you have a place were one rule's out the other.

They say that the centre of the sun and planets are a Solid and it is iron. The core of these objects are at their hottest So why are they still in a Solid state. And if it was a sun and as my observations pan out if they were a Solid when the sun exploded they would not have burst out into our universe. This is based on Isaac Newton's theory the force is equal from each side or in this instance sides So that means that they would more likely be where they are and there would not be a Black Hole. I think that these physicist and scientists are trying to out do each other with their theories and each one gets more preposterous. Gases when they reach a temperature that would cause them to explode and this happens I would think they would turn into another gas

STELLAR REMNANTS

Stellar remnants is the name they give to the death of the star (when a star has burned out its fuel supply), it will undergo a contraction that can be halted only if it reaches a new state of equilibrium. Depending on the mass during its lifetime, these stellar remnants can take one of three forms:

White dwarfs, in which gravity is opposed by electron degeneracy pressure[3]

Neutron stars, in which gravity is opposed by neutron degeneracy pressure and short-range repulsive neutron–neutron interactions mediated by the strong force

Black hole, in which there is no force strong enough to resist gravitational collapse

White dwarf

The collapse of the stellar core to a white dwarf takes place over tens of thousands of years, while the star blows off its outer envelope to form a planetary nebula. If it has a companion star, a white dwarf-sized object can accrete matter from the companion star until it reaches the Chandrasekhar limit (about one and a half times the mass of our Sun) at which point gravitational collapse takes over again. While it might seem that the white dwarf might collapse to the next stage (neutron star), they instead undergo runaway carbon fusion, blowing completely apart in a Type Ia supernova.

NEUTRON STAR

Neutron stars are formed by gravitational collapse of the cores of larger stars, and are the remnant of other types of supernova. They are

So compact that a Newtonian description is inadequate for an accurate treatment, which requires the use of Einstein's general relativity.

WRONG

If I have said it once I'll say it again in case Some of my readers have skipped a page. Nothing in space or have I have said this applies to earth, can implode.

LOGARITHMIC

plot of mass against mean density (with Solar values as origin) showing possible kinds of stellar equilibrium state. For a configuration in the shaded region, beyond the black hole limit line, no equilibrium is possible, So a runaway collapse will be inevitable.

According to Einstein's theory, for even larger stars, above the Landau-Oppenheimer-Volkoff limit, Also known as the Tolman–Oppenheimer–Volkoff limit (roughly double the mass of our Sun) no known form of cold matter can provide the force needed to oppose gravity in a new dynamical equilibrium. Hence, the collapse continues with nothing to stop it.

Once a body collapses to within its Schwarzschild radius it forms what is called a black hole, meaning a space-time region from which not even light can escape. It follows from a theorem of Roger Penrose[5] that the subsequent formation of Some kind of singularity is inevitable. Nevertheless, according to Penrose's cosmic cenSorship hypothesis, the singularity will be confined within the event horizon bounding the black hole, So the space-time region outside will still have a well behaved geometry, with strong but finite curvature, that is expected[6] to evolve towards a rather simple form describable by the historic Schwarzschild metric in the spherical limit and by the more recently discovered Kerr metric if angular momentum is present.

On the other hand, the nature of the kind of singularity to be expected inside a black hole remains rather controversial. According to Some theories, at a later stage, the collapsing object will reach the maximum possible energy density for a certain volume of space or the Planck density (as there is nothing that can stop it). This is when the known laws of gravity cease to be valid.[7] There are competing theories as to what occurs at this point, but it can no longer really be considered gravitational collapse at that stage.[8]

Theoretical minimum radius for a star

The radii of larger mass neutron stars (about 2.0 Solar mass) are estimated to be about 12-km, or approximately 2.0 times their equivalent Schwarzschild radius.

It might be thought that a sufficiently massive neutron star could exist within its Schwarzschild radius (1.0 SR) and appear like a black hole without having all the mass compressed to a singularity at the center; however, this is probably incorrect. Within the event horizon, matter would must move outward faster than the speed of light in order to remain stable and avoid collapsing to the center. No physical force therefore can prevent a star smaller than 1.0 SR from collapsing to a singularity (at least within the currently accepted framework of general relativity; this doesn't hold for the Einstein–Yang–Mills–Dirac system). A model for nonspherical collapse in general relativity with emission of matter and gravitational waves has been presented.

The astrologers have come up with the notion that black holes appear when a sun that is twenty times more massive than our sun and it collapses into itself. How can we as mere mortals even begin to understand this?

I hope that I am not the only one that has just had their mind collapse into a black hole from the gobbly gook in the above theories given above. I will take it at a piece at a time. A black hole comes about when a star gets So hot and uses all its energy it collapses. This is because the gravitational force is not stronger than the force of the collapse have you got that. Maybe there is Someone out there could explain to me when one minute the physicists, and scientist say that when a sun collapses and creates a black hole the gravity is So strong that not even light can escape from it. Now they have just said that the gravity when a sun collapses is stronger than the gravity before the collapse. which one is it? I am lost. The stars that are collapsing seem to get bigger and bigger. We know that man is inclined to exaggerate So take this with a pinch of salt. You know my opinion of black holes, So I won't go into them details again. How about this for a question? We have had the Big Bang, this gave us the universe. And the start of this was the Milky Way that was the start of the first generation of galaxies, as each sun in the these galaxies were of a different size So the demise of these sun's accrued at different times. Each of these bangs created a black hole, and each black hole had quite several galaxies spinning around it. And all of these new galaxies like everything else in the universe was travelling in an out ward direction. Nothing was collapsing So what was it that caused this? There is no sign from the latest photos from NASA or any other body taking images of space that there was anything to show that the universe was in fact getting smaller. The matter that is left after the Big Bang gets smaller and smaller and travelling further and further away. There will come a time when it is that small that our sun's will not be able to detect them. And that in fact will give the physicists and scientists that their theory about the universe collapsing on its self was correct.

They say that there is a black hole in the Milky Way. Let us presume that the Big Bang happened and it was that, that created the Milky Way. It has been noted there is a black hole in the Milky Way. I can say without any contradiction, (I hope,) that there is a black hole at the centre of every galaxy. The size of the black hole in the Milky Way is four million times the mass of our sun. Now that's big. It is Also surrounded by gas. Imagine matter packed So densely that nothing can escape. Not a moon, not a planet and not even light. That's what black holes are — a spot where gravity's pull is huge, ending up being dangerous for anything that accidentally strays by. But how did black holes come to be, and why are they important? Below we have 10 facts about black holes — just a few titbits about these fascinating theories.

MY OBSERVATION

Let us look into a black hole. Not to close I cannot afford to lose any of my readers.

Every time a sun exploded it creates a black hole. You might ask how I came to that conclusion. All the contents of a sun when blasted into space leave a black hole. What this amounts to is a void where there is no matter of any type, when you have this, what have you got. There is nothing that can reflect the light. Then surly there is more sense in this than its gravitational is that strong that light cannot escape from it. It does not matter whether it is gases, Solids, liquid or dust. Now let's look into this, no pun intended. Now I have a real problem with the theory that a black hole gravitational swallows everything that surrounds it. If this is So then that's the end of our universe as we know it. The last time I looked the universe was still here So that theory can't be right. As I have pointed out earlier that gravity holds the planets and moons in an orbit around its sun. Gravity does not pull objects to the ground. it is the weight of the object that makes it fall. Why do two objects of a different weight take the same time to reach the ground? This is Isaac Newton's finding of equal forces balance things equally. This Also applies to a certain degree on the way galaxies work, sand what influence they have on the gases in the universe.

MIGRATION

Let's have a look how fish, birds and insects etcetera find their way South. The naturalists say they are being directed by the moon. Others say it's the magnetic pull. These two theories do not put into the equation that none of these tell them when it is time to start on their journey the moon and the magnetic force are always there. So to me this is not the answer. So what other explanation can there be. All things are governed by nature or by the rules that make the planet tick.

The miracles that were performed in the bible can all be explained that this was possible by the acts of nature.

As I point out in this book about God and the Garden of Eden what could direct them South is the wind. As the weather gets colder then they would be able to feel the difference in the drop in the temperature. When they got a whiff of the warm air they decided it was time to move house. Let us put ourselves in the world of the birds. I am going to the extent of us flying South. When the Earth was of one land mass and the birds flew South. This is quite reasonable to understand. Now I would like to know? How did the animals cope with this problem? Going by the geologists the reason that the land mass gradually parted, Sorry this is the bibles account.

The geologists account is that it was through erosion. If this is the case, then at what time did the animals realise that the distance that they had to swim was too far and they would have drowned through exhaustion. Maybe when they got half way they turned back, it's a joke. This is not as stupid as it seems the wilder beast cross alligator rivers on their movement going north to South and vice versa. Are their movements controlled by the moon, and if not why not. Maybe as their field is one that covers thousands of miles they look around and see that the grass is greener in one direction rather than another. Do the birds see this? I am trying to put insets, birds and animals in the same category. Of course not there is only one thing that can make all living things react to Something and that is the wind. When you see living creatures, and I should imagine it applies to insects. When there is cold wind they all ways have the wind at their back. That should bring you to the decision that as the spring/summer approaches they feel warm air on their fronts. And of they go.

Let us take a look at what happens to the British Isles. The land is gradually falling into the sea. And it is slowly ground down into sand. Now why is the seabed not all just one massive sand beach? I know that there are parts of the sea bed have mountains, now we know that they are formed by volcanic eruptions and quakes, we Also know that these accrue at the faults of the Earth's plates.

had no reason to fly South, all that they needed was a place that was environmental friendly.

How is this for a theory? Maybe this is how Wales, fish. Find there way around the oceans. Let's take salmon and sardines. Taking the salmon as are first adventure the naturalists say they can remember where they were spawned. There's a lot to remember I'm glad I'm not a salmon even with my computer brain I'm always getting lost. And I'm not the only one. So they must be another reason why they can find there way. The only logic thing that it can be is the currants. They would wait until the currents got warm and then they would follow them. These would have the same effect on fish as winds had on birds. The currants must Also have different temperatures. When they come from the South they must be warmer. The sardines Also migrate as this can be seen on wild life TV. Programmes they Also follow the warm currants and the predators must Also use this method that is why they know that dinner will be arriving. They don't have a radar system. I believe my theory must more reliable than all the other theories. This is a more logic reason for birds and fish to find there way. Let's put ourselves in the bird's environment. The kids have flown from the nest to find a life of there own. Well dear you say to your mate. It's very quite since the chirping has stopped I don't know what we are going to do, and were not going to your mothers. You're sitting there looking at the moon you turn to your better half and say. 'I am not getting any messages from that how about you?'

'Your right there, I have heard the humans say that we are guided by the moon to move South. I've had a good look and there's no weather cock on it to say which way is South. Have you any idea which way it is?' 'No I've not a clue. Maybe if we sit quietly we might hear Something.' the night passed, and the moon just hung there. The next day dawned. It had been a little colder during the night. And the sun was a lot cooler the next day. 'We can't stop here you say to your feathered companion. Do you remember last year we did go South.' there must have been Something that got us on the right track. Then out of the blue they feel a warm breeze on their face. 'That's it, it is the warm wind that's the way we must go, follow the warm wind. It's a good job we are not relying on the moon. I mean what do we do of a day time when there is no moon, sit around on are feathers until night time. Look there's the crowd that we were with last year. Let's go.' I know there bird brained but this is much better that the moon theory.

How long did it take for dogs and cats to realise that vegetables were good for them. Wild dogs and cats don't eat vegetables. So why are we giving dogs and cats vegetables in their food.

Garden of Eden. There were no insects, grubs or weeds in the Garden of Eden So how did the plants get pollinated.

Earth as only one land mass
                    this is not possible read on.

Let's must look at Earth's if there was one land mass. Our planet is 75% water and using my mental arithmetic I have worked out that the land mass is 25%. We now have this big chunk of land. And seeing that the Earth's land mass became about because of the planets volcanic terrain. The next thing we need to transform this land is rain and wind. Another thing is defiantly vegetation; otherwise in no time there would be no land left. But we can disregard the vegetation and every other living thing. The rain seems to gorge out deep canyons. These are more noticeable in dry regions. There would be nothing to protect the Soil So that would be washed away. This then would cause the rocks to plunge down the mountain side. We know there are faults to the Earth's crust. And we know that they are positioned over a large area. That means that the majority of these faults are under the sea. But we do not have any sign of them remember Earth has only one land mass. This to me seems very strange. But let us continue. With there being no protective covering on the land as in vegetation the land will decrease in its mass at an alarming rate. As the rivers form they would be a lot wider but shallower owing to the amount of silt. Don't let us forget there would be a minimum of land mass created by the land volcanoes. The land would become flatter and finish up as deserts. We can see at today's erosion how the weather can destroy land mass. Just look at our coast line and the way estuaries keep filling up. And we have vegetation to protect it. Can you imagine how the land would have fared over just one billion years? Here's a good point to consider. We have are four continents. Just think about it. But they all belong to one land mass. And these were created by one land mass being eroded over billions of years. Think about it. I must put my thinking cap on and edit what I know and study the rest. Studying yuk.

I will start with what I know. The distance between Europe and the American continent is two thousand miles. You might think that this is a lot of corrosion.

How about Australia.

Does anyone on this planet can honestly say

The astrologer's scientists and physicists keep moving the goal posts back. When they got to are nearest planet, the moon and found no little green men. They then said that there is life on are next nearest planet mars, but that was too hot. Here on Earth there are two extremes in the temperature. Let's take a stroll in the desert. First put on your sun block, Bermuda shorts, and maybe an umbrella. a flask may be a good thing to have. I nearly forgot sun-glasses. They then came up with a brilliant idea that they could live beneath the surface. Don't get me wrong this is plausible as there are creatures, insects and water creatures that do live on our beautiful planet Earth that live below the surface. There are fish that burrow into the sand and lots of other living things that Also hibernate in wonderful ways. They can survive many years in this state waiting for the rains to come. Some periods the rain does not come for seven years and probably longer. We cannot see them but they are there. So it's very probable that Mars has the same living things below their surface. We can see there are cacti, So plants can Also survive in these extreme conditions. There is another point they say that there could be below the surface on Mars. So this looks to me like there is a great possibility that they might be living under ground. That should Also there are little green men. There more than likely be a white sickly colour. Now all these living things are waiting for the rains to come but there are no rain clouds. In time I mean billions of years any water under the ground must have evaporated. Where is it?

Now we will take a tour around the Antarctic. There is moth that when the winter comes it goes into a chrysalis this then freezes Solid. I mean if you used a hammer and chisel you would have a job to break it open. If they can stand the temperature, that goes down to fifty degrees below freezing and in our deserts at one hundred Fahrenheit, and then forty degrees below at night what does this tell you. All the things that live on the Earth should be able to live on other planets. So why is there nothing there? Let's take the view that there is a God and he created the Earth and all living things. Why did he not create living things on the moon that really would confuse our scientists?

Let's voyage to the next planet Jupiter. Let us take a stroll along Jupiter's surface this planet is cold we will need are thermals on and a pair of ear muffs. Thermal gloves a hooded fleece and a warm pair of shoes. I nearly forgot a packet of tissues.

If the moons gravity is not as strong as Earth's then physics can surly see that it cannot affect Earth's gravity, if this is the case then it cannot have any effect on Earth's tides. Or any other thing on our planet whether it be animal vegetable or mineral. I mean what are the elements in the water that can be affected by gravity. I mean there must be Something. The boffins have not said how this is happening. What we have is nothing pulling nothing. That's clever.

Everything that is happening now in space is just a repeat of the Big Bang that happened billions of years ago. There is nothing to discover that is new no matter what planet you go to. They have built a long tunnel to register any pulses from outer space. This has taken ten years to build and they have received three pulses or using another word quivers. UREACA I don't know how much it has cost, but was it worth it to mankind. Has anything been of any value in all this time since they first started exploring space. I can supply them with quivers I got them when I thought about the amount of tax payer's money it has cost us. I could have given them Some pulses with a lump hammer.

THE BIBLE

Why can you not trust what is written in the bible

Answers to the bible

I am now going into the world of our creator, and the readings of the bible. All over the world every country or tribes have their own Gods. You can go to places that are So remote and have no knowledge of the way us Europeans, Asians and other civilised people worship our God. They may worship the sun god or other gods in the heavens and things of the Earth. Let us look at the way we worship. All the different races have there own name for their God. What we need to understand is the basics of the bible. There is the story of the beginning. There is the creation of the universe, we all now about this as it is all around us. We can go along with the astrologists and scientists that the Big Bang was the start of our universe. As can be noticed when you read the bible there are lots of contradictions. Which ones are you supposed to take at there face value. The bible is not to be trusted, and why is this. There are So many contradictions. Let's look at Some of them. An eye for an eye, and a tooth for a tooth. Contradiction. If you are struck across the cheek turn the other cheek. What does this tell you? It is what the cover said, the bible cannot be trusted. These are traditions that all the religions have and they are there basically to control their Society. It has nothing to do with the word of the Lord. This covers what you cannot eat, bacon you cannot eat beef you cannot eat meat on a Saturday/Sunday. What they are doing is brainwashing their Society to believe that this is the word of the Lord. any Society that condemns people in their jurisdiction from having a free mind this is a case of impriSonment of the mind. When you get a Society that is using this kind of behaviour and they are not stopped then you get a leadership that goes down the road to a Dictatorship or worse still an ISA terrorist group.

The reason there is So much contradiction, is that the bible was written by humans. But like all human beings they are likely to exaggerate. They will Also put in their own interpretations and they will embroider the facts. That is why their observations must be taken with a pinch of salts. All religions have made up laws by males and that is why they bring out laws that keep woman in there place and to make sure that woman understand that their religion has deemed this to be. You don't need all these books on religion So lets get rid of all of them.

There is only one part of the bible you can trust in and that as you know is from the very beginning. We know by D.N.A. results that what the bible says is right when it says, there was Adam and Eve. We Also know that the family consisted of two brothers and five daughters. This has been proven by D.N.A. results. D.N.A. results say that there are six families in Europe. As most of the world has seen Europeans migrating to these lands we can see why this is the case. There has been tests done on inhabitant that have not come from Europe and their D.N.A. matches to one of these Families. You might have done Some working out for your self and come to the conclusion that my sums don't work out. I said there were two males and five daughters that make seven families. If you know your bible then know it says that Cain killed Abel. Bingo six families. Let's go further on as I have already written about how the other males and females came on the scene. We must be selective about what we are going to believe in. now we have Noah. There have been no findings concerning this part of the bible So we will must ignore it. The next one concerns Sodom and Gomorra. There has been proof of this by archaeologists they have found five sites that appears to have been cites that are covered in Soot. There were a number of cities that were not destroyed. This seems to be against nature's way of going about things. God told Job to collect all his people together and leave the town of Sodom, as he was going to destroy it and all that lived there. How it was possible for four cities to burn down at the same time, as the bible had said this would happen? The remains of these cities can be seen that the rocks turned into dust/ash. To do this the temperature had to be 6,000 degrees Fahrenheit I hope any readers that are

Bang Goes That Theory

new to the bible's revelations will bear with me concerning the information below and can find Something of interest in them.

Sodom and Gomorrah

(/ˈsɒd.əm/; /ɡəˈmɔːr.ə/[1]) were cities mentioned in the Book of Genesis and throughout the Hebrew Bible, the New Testament and in the deuterocanonical books, as well as in the Quran and the hadith.

According to the Torah, the kingdoms of Sodom and Gomorrah were allied with the cities of Admah, Zeboim and Bela. These five cities, Also known as the "cities of the plain", (from Genesis in the Authorized Version) were situated on the Jordan River plain in the Southern region of the land of Canaan. The plain, which corresponds to the area just north of the modern-day Dead Sea, was compared to the garden of Eden as being well-watered and green, suitable for grazing livestock.

Divine judgment by God was then passed upon Sodom and Gomorrah and two neighboring cities, which were completely consumed by fire and brimstone. Neighboring Zoar (Bela) was the only city to be spared. In Abrahamic religions, Sodom and Gomorrah have become synonymous with impenitent sin, and their fall with a proverbial manifestation of divine retribution. Sodom and Gomorrah have been used as metaphors for vice and homosexuality viewed as a deviation. The story has therefore given rise to words in several languages. These include the English word *Sodomy*, used in Sodomy laws to describe sexual "crimes against nature", namely anal or oral sex (particularly homosexual), or bestiality. Some Islamic Societies incorporate punishments asSociated with Sodom and Gomorrah into sharia.

ETYMOLOGY

The etymology of both names is uncertain. The exact original meanings of the names are Also uncertain. The name *Sodom* (Hebrew: סְדֹם *Səḏōm*) could be a word from an early Semitic language ultimately related to the Arabic *sadama*, meaning "fasten", "fortify", "strengthen", and Gomorrah (Hebrew: עֲמֹרָה *ʿĂmōrāh*) could be based on the root *gh m r*, which means "be deep", "copious (water)

There are Some other stories and historical names which bear a resemblance to the Biblical stories of Sodom and Gomorrah, and Some possible natural explanations for the events described have been proposed, but no widely accepted or strongly verified sites for the cities have been found. Of the five "cities of the plain", only Bela, modern Zoara, is securely identified, and remained a settlement long after the biblical period.

The ancient Greek historiographer Strabo states that locals living near Moasada (as opposed to Masada) say that "there were once thirteen inhabited cities in that region of which Sodom was the metropolis". Strabo identifies a limestone and salt hill at the South western tip of the Dead Sea, and Kharbet Usdum (Hebrew: הר סדום, *Har Sedom* or Arabic: جبل السدوم, *Jabal(u) 'ssudūm*) ruins nearby as the site of biblical Sodom.

This occurrence was brought upon man because of one of two reasons when man was to feel the wrath of the lord. The other reason was when man worshipped false gods. These do longer apply. When Jesus was crucified all mans sins were forgiven. There was a time in the life of Jesus when a stranger arrived in the city and there was a crowd of men waiting outside the dwelling where this stranger was staying. Jesus approached them and said why do you wait for this stranger to appear, he is a guest to you city. This was to commit a homosexual act. You will note he did not chastise them for their behaviour only the way they where using mob rule to have their way with the stranger. There are certain traits in the human building blocks that we can overcome, and there are other traits I find must be hard to over come. Do not get me wrong were there is a case that a perSon finds it hard to combat this crime then he should not get into the circumstances that he likely to commit this crime. If a perSon has not committed that crime for a long period of time then surly we should forgive him. If he does commit the same crime then the right to his freedom must be curtailed. I hope I have not gone to far from the subject but I think that there is a connection.

Have you thought what the one-thing in the bible that you can go by? Forget all that you have read or been told. I will give you a clue, Ten. It is of course the Ten Commandments. When you think about these you will realise that if man kept to them we would paradise here on earth. This could never happen. Why I hear you ask? The reason is because Some people have mental problems. Of all the things that should not be this is the main one. We must understand when our Lord created man and man disappointed the Lord he wrecked vengeance upon man. It was only when his Son was crucified and he asked his father to forgive man that all men could enter the kingdom of God. We all know that God could not be of human form and he would never show himself this way. I have mentioned before that was because no matter what form he showed himself he knew that not all the humans would like that image. There is Something I cannot get my head around and that is fore Some one that had no sympathy for others could be So upset by what his creation might think of him in human form. That's my lesSon for today I hope you can understand why I had to get it out of my system.

Einstein did an experiment were he used a prism to remove the colours from the sunlight. As I have said I am trying to use logic instead of theories. Going by this experiment it looks like and does show the colours of the rainbow. I have never done any experiments on this scenario. To me Something doesn't seem right. He points out that this proves that there is no white light but the colours mixed together produces a white light. I can't get my brain around this, So I started to think.

There was three lasers using the colours blue green and red, this showed a light being shone on a screen. Each one showed the colours of each laser. I noticed the centre of each one was white. If the combined colours show white why does each one show a white centre? What this shows is that the white light is stronger than the colours.

The invention of their new analytical technique, spectroscopy, can lead to the discovery of many new chemical elements in the years ahead.

Kirchhoff was interested in the spectrum of sunlight and realized that Fraunhofer's black lines corresponded exactly to the bright rays emitted by specific chemical elements. He deduced that the white light produced by the hot surface of the Sun was in part abSorbed by certain chemicals all these experiments are done in our atmosphere. They say that this is what is happening on the sun's surface.

The invention of their new analytical technique, spectroscopy, led to the discovery of many new chemical elements in the years ahead.

Kirchhoff was interested in the spectrum of sunlight and realized that Fraunhofer's black lines corresponded exactly to the bright rays emitted by specific chemical elements. He deduced that the white light produced by the hot surface of the Sun was in part abSorbed by certain chemical elements present in its cooler atmosphere, which produces black lines in the spectrum. In fig three the black line only show as slits of black

The geologists and scientists say that the reason the sea is blue is not by the reflection of the sky but by the elements in the sea that reflect the light into our eyes. Now my question is if this is the case then have these elements disappeared in the salt flats at Utah, and what would have caused this, and how about this if that is the case does ever thing that is in the universe has elements in them that shows their colour or is it the white sunlight that shows their colours. Let's look at this logically. Every thing has colour pigments. There can be seen with every thing around you. It is in the flowers, rocks or the material. Let us take the scenario that the scientist says that then reason that the sea is blue is because of the pigments in the sea. I have seen the sea on a dark day the sea has suddenly lost its pigments; it's no longer blue the sunlight is still getting through. So what is happening? So it's a cloudy day and there is no direct sunlight. The above diagrams show the seven colours of the sun's rays. Then by using a computer to blend the colours together they get a white glow. They do more computerising and they get black lines. do the scientists know that with a computer we have all strange thing. How about superman, that's a start I could go on with monsters and lots of other things, but you get the gist of my thinking. the point I am going to put forward is. How can you mix light colours to together and get white and black. why does a prism show these colours. The colours are the result of gases. Different gases different colours. A prism or a mirror cannot move gases what they do is a reflection of these colours. You can see these colours around the circumference of sun's and Also planets. There is no prism that allows these colours to be seen. The sun consists of atoms and particles. The colours do not mingle to make a white light what accrues is the colours are there but the white rays are much stronger that the colours cannot be seen. Now we will go into the reason why we can see the colours at the circumference. We are looking at them at an angle that is more condensed. As the Earth moves around the sun So the gases seem to move this does not affect the colours. We should not be able to see any of the colours it they change into a white light. The gases being lighter in weight can be seen on the circumferences of the planets and sun's. As we know by the way things work here on Earth, I always come back to this way of thinking as most things can be worked out this way. My point is in factories where a lot of heat is used as in the steel industries the hotter the furnaces gets then it gets white hot. You will not see any colours but they are there. And that is why you only see the colours at the edges of anything that produces heat. If you think about it this is why you do not see the colours from the centre of the sun.

Please note I have not done any experiments on colours and gases I wouldn't know where to start. I am working on logic only.

Prism. The reason only the colours can be seen when you pass the sun's rays through a slit is because you are reducing the strength of the white light. And this is only a reflection that shows the colours

??? as with Some planets there is an atmosphere, any gases that can be seen from these planets define the metals and other material that are on that planet Our moon has no gases So no colours. Our Earth has an ozone layer that creates a blue colour. Below this ozone there are gases being produced by the different metals maybe just maybe the reason we see all these colours when a prism is used to separate them from the sun's white colour is not because they are from any colours from the light from the sun but they are there because of the gases/colours.

As things evolve there seems to me that all these species seem to be the advancement in their size. They get bigger and they adjust to there environment. So why have we the human species have got smaller. This is in compariSon to the big apes. There all are different ape families in different parts of the world they are all bigger. But it is only in the last fifty years that the human race height has increased by three inches. This is due to medical enhancement.

Everybody knows you can't produce Something from nothing. it does not matter if it is animal vegetable or mineral As different species advanced they adjusted to live to their environment as did all the creatures, insects, and birds. If every-thing evolves here's a question? You have a group of insects, animals or birds. They would all must advance other-wise the ones that didn't evolve then surly they would not be able to survive. That tells me that the strong will survive and the weak perish. So if this is the case then the original species would become extinct. The experts say we originated from the sea. We were just microbes. These microbes lived on plant life. They must have got bigger I mean just take a look at elephants. They could not live on any other form of life as there was none. So how could they evolve any further? That's it. Let the experts tell us what happened. You will find out at a later date why everything scientific proves that we never came from apes. I know that all children like to climb trees, this is not that they are going back to there ape ancestors. If this were So then we must be going back as fish as we love the water. We do these things for fun it has nothing to do with are past

Going back to the weight of material that the physicists say that on the gravitational pull is So strong that a spoonful of material from this planet weights millions of tonnes. Now, how can anyone in there right mind believe this. Every thing in the universe is Also apparent on Earth.

And nearly all that happens in the universe has happened on Earth. Maybe not in the same proportional but near enough So us mere mortal's can understand. *Gravity does not compress anything.* To me that must mean that they are using this word to state a fact. I have done away with the term gravity instead we should use my word. Kane Force.

How about trying to compress a jelly. Do you think there is anyway that you can compress a jelly? No So what happens. It just spreads out. Lets try Some thing else. We can put the jelly in a container that is sealed. We press the plunger, if there is no were that the jelly can expand to as in a sealed container does the jelly compress into a smaller jelly and that makes it heavier, no it does not. no matter what you compress whether it is liquid gas or a Solid, it compress into a smaller amount, but once all the air in these has been expelled than it cannot reduce in size any more.

This applies when the said substance is in a confined space. If it is not in a confined space then it will expand at the weakest point. This then what would happen on the planets as I have said I am no expert in anything concerning space, as in physics astrology, or science? Why I became interested was the theories that was being put foreword by the said experts was So ridiculous that I started to look at things logically. Sorry I am digressing. How did I come to the logic decision that gravity does not draw things in, the universe is expanding, I looked at the reasons the astrologists, scientists and physicists gave and I could see that there research and the time table that photographs of the universe were taken at different time of the year and it showed that this was So. The time scale of this expansion was calculated it seventeen thousand miles per second. This made me come to the conclusion that gravity had no effect on how any thing on any planet is affected by this phenomenon. What gravity does do it moves all matter in the universe outward and then the universe takes it in a circular direction. What would have happened if this was not So. When the Big Bang accrued everything would have carried on in an outward direction. As there was no sun's gravity then the matter caused by the Big Bang would mean that they would be travelling in an outward direction. That would have meant that the universe would have a different outlook that it has now. There would be no Solar system as there is now.

Let's look at it from another prospect. There would be no moons going around planets no gases would be seen, and there would be no objects in space hitting any other planet. And as the sun's created by the Big Bang would not be pulling their moons and planets around them. This in effect would mean that they would cool down a lot faster. And they would all finish up as frozen planets. Why because spaces temperature is absently zero. This then would cause the heat to be diminished. In time when this zero temperature started to cool the sun's heat at its surface then a crust would form. And in the end it would be as the Earth is. There would be Earth-quakes and volcanoes until the crust was that thick the sun would explode. The bible forecasts that God will intervene when the Earth is rained upon with fire and brimstone, So preventing an explosion. This is I know very hard for the non believers to understand but later I will explain beyond a doubt why this is So.

Researchers believe they have found evidence of real natural disasters on which the ten plagues of Egypt, which led to Moses freeing the Israelites from slavery in the Book of Exodus in the Bible, were based.

But rather than explaining them as the wrathful act of a vengeful God, the scientists claim the plagues can be attributed to a chain of natural phenomena triggered by changes in the climate and environmental disasters that happened hundreds of miles away.

Archaeologists now widely believe the plagues occurred at an ancient city of Pi-Rameses on the Nile Delta, which was the capital of Egypt during the reign of Pharaoh Rameses the Second, who ruled between 1279BC and 1213BC.

The city appears to have been abandoned around 3,000 years ago and scientists claim the plagues could offer an explanation.

Climatologists studying the ancient climate at the time have discovered a dramatic shift in the climate in the area occurred towards the end of Rameses the Second's reign.

By studying stalagmites in Egyptian caves they have been able to rebuild a record of the weather patterns using traces of radioactive elements contained within the rock.

They found that Rameses reign coincided with a warm, wet climate, but then the climate switched to a dry period.

ProfesSor Augusto Magini, a paleoclimatologist at Heidelberg University's institute for environmental physics, said: "Pharaoh Rameses II reigned during a very favourable climatic period.

"There was plenty of rain and his country flourished. However, this wet period only lasted a few decades. After Rameses' reign, the climate curve goes sharply downwards.

"There is a dry period which would certainly have had serious consequences."

The scientists believe this switch in the climate was the trigger for the first of the plagues.

The rising temperatures could have caused the river Nile to dry up, turning the fast flowing river that was Egypt's lifeline into a slow moving and muddy watercourse.

These conditions would have been perfect for the arrival of the first plague, which in the Bible is described as the Nile turning to blood.

Dr Stephan Pflugmacher, a biologist at the Leibniz Institute for Water Ecology and Inland Fisheries in Berlin, believes this description could have been the result of toxic fresh water algae.

He said the bacterium, known as Burgundy Blood algae or *Oscillatoria rubescens*, is known to have existed 3,000 years ago and still causes similar effects today.

He said: "It multiplies massively in slow-moving warm waters with high levels of nutrition. And as it dies, it stains the water red."

The scientists Also claim the arrival of this algae set in motion the events that led to the second, third and forth plagues – frogs, lice and flies.

Frogs development from tadpoles into fully formed adults is governed by hormones that can speed up their development in times of stress.

The arrival of the toxic algae would have triggered such a transformation and forced the frogs to leave the water where they lived. Knowing a bit about nature from watching David Attenbougher's nature films, and using a bit of common knowledge, from being a gardener. There is one thing that covers all living things, this includes plants. Every thing is governed by their food supply in their environment. Also what is more important is the water supply. We now that the animal kingdom can last longer in there environment longer than three days. But this is most important, the temperature had risen belong belief.

If there is a shortage of food then the young are the first to die. We will go back to the frogs. The forecast given by Moses was a plague of frogs. These frogs were everywhere. In their kitchens, bedrooms, in fact every step was followed by a crunching Sound. So before the river was turned red by the algae there had to be a billions of tadpoles. These all turned into frogs. Moses had seen all these tadpoles/frogs when he was walking by the river side. It looks like we are going to have a plague of frogs. I will inform the Pharaoh that God would send a plague of frogs that should scare the sh…sh…sh. shoes off him, or sandals.

The next plague was lice. This would be a natural thing. As the frogs died there would be no blood for the lice to live on So they would depart from their dining room table. God came to Moses and told him to tell @@@??strike the ground with his staff, and this brought forward the plague of lice. Now we have decaying frogs and lice. There was still no surrender by the pharaoh. From the dead frogs we have the flies. We can all see this natural thing taking place in order of circumstances. The natural accuracy from the dead bodies would be maggots. These in turn would turn into flies.

Now we have the mosquitoes. What we know about mosquitoes is that they Also are governed by the hot weather. And what we do know is that they carry germs, diseases, including malaria.

These would have flourished without the predators to keep their numbers under control.

This, according to the scientists, could have led in turn to the fifth and sixth plagues – diseased livestock and boils

ProfesSor Werner Kloas, a biologist at the Leibniz Institute, said: "We know insects often carry diseases like malaria, So the next step in the chain reaction is the outbreak of epidemics, causing the human population to fall ill."

Another major natural disaster more than 400 miles away is now Also thought to be responsible for triggering the seventh, eighth and ninth plagues that bring hail, locusts and darkness to Egypt.

One of the biggest volcanic eruptions in human history occurred when Thera, a volcano that was part of the Mediterranean islands of Santorini, just north of Crete, exploded around 3,500 year ago, spewing billions of tons of volcanic ash into the atmosphere.

Nadine von Blohm, from the Institute for Atmospheric Physics in Germany, has been conducting experiments on how hailstorms form and believes that the volcanic ash could have clashed with thunderstorms above Egypt to produce dramatic hail storms.

Dr Siro Trevisanato, a Canadian biologist who has written a book about the plagues, said the locusts could Also be explained by the volcanic fall out from the ash.

He said: "The ash fall out caused weather anomalies, which translate into higher precipitations, higher humidity. And that's exactly what fosters the presence of the locusts."

The volcanic ash could Also have blocked out the sunlight causing the stories of a plague of darkness.

Hang on a minute there seems to be a clash of scientist's findings. They have said that the ash had made the temperature rise. So this would be exactable what was needed to start the locus outbreak. Have you seen the trouble with this scenario? It contradicts the way nature reacts to the weather. I agree with the heat this would definitely cause the locus to start to react and start coming to life from plasma to there maturity as locus. If you have not seen what is wrong with this scenario is, the ash blocked out the sun this caused. The darkness was that dark you couldn't see an elephant in a broom cupboard. I am confused, what came first. The scientists say that the hail came first. What started it all as found out by the scientists was the drought. To get back to what caused the hail? If it is as they say then the ash came and this clashed with the thunder storms in turn produced the hail. So is this the end of the drought, and how long would it last? Let's bring forward the locus. These must have came from a different area because otherwise the hail would surly have killed them or a severe case of headaches. And surly this would have brought on the darkness. This seems a bit mixed up to me but I am not here to Sort out the scientists problem let them Sort it out. Also if they were having thunder storms and hail then surly all the dangerous acids would come down to Earth. This would have been the end of the Egyptians. There would then be no need for the exodus of the Jews, they could have stayed where they were and became rulers of

This was because it must be the locus, how have I come to this decision. The locus might have evolved from alga but there is now way that they were going anywhere. That meant they had to wait until the darkness ceased to be.

Scientists have found pumice, stone made from cooled volcanic plasma, during excavations of Egyptian ruins despite there not being any volcanoes in Egypt.

Analysis of the rock shows that it came from the Santorini volcano, providing physical evidence that the ash fallout from the eruption at Santorini reached Egyptian shores. All the forecasts predicted by Moses concerning the ten plagues can be explained by naturalists and scientists. Let's take the first one all the waters turning red. This was probably caused by a volcano erupting, and the dust content mainly being of sulphur. So when it erupted the wind had to be blowing in the direction of Egypt. Now I have a problem. If the river was beyond Egypt then Egypt would Also be covered in sulphur. If the river was before Egypt, then when it reached the river did the wind change, or did it cease to blow. Wouldn't the sulphur being a strong poiSon have killed every one? Or they would have had a serious case of blisters. Let's presume that the volcanic dust came quickly, and this would certainly make the frogs leave the river right away. *Clever frogs* they can move faster than the wind. There would be no food for these frogs. So how long would it take for them to die? They would have no protection from the sun, let's say two days. The frogs hopped into the happy ponds in the sky. The decomposing bodies rotted, and So we have the plague of lice. These would certainly not last long with no blood for dinner. One day. That's a total of four days. Now we have the flies. God said to Moses tell Aaron to strike the ground with your staff, and all the dust will be flies, lice, and mosquitoes.

There is a point that I would like to make. With the start of the plagues whether it is from the reaction of the alga starting the plagues or the volcano there effect would be the same. But they would be at different times of plagues.

The Jews food store was not affected by this plague as our Lord did not put a curse on the Israelites

The Jews sacrificed the sacred lamb, and put a sign with its blood on their place, this was to show God that it was a Jewish abode.

We shall look at the first one. Volcanoes. This accrued 400 miles away. How could Moses know this would happen when he predicted that the water would turn into blood? Also that the wind would be So favourable, that it would blow in the right direction. Remember that over 50 days there were ten plagues. This is an average of five days per plague.

We have all heard of the natural elements that could have caused the ten plagues of Egypt. All the miracles that accrued in the bible can be explained that there was a natural reason for these happenings. God used all the elements that he created to do these miracles. Scientists look at the way the rocks have left there impression on what happened in the past. We will look at each this as we go along. Moses' was not only Chosen by God to do his biddings but Moses could surly see into the future; he didn't go digging amongst the rocks.

I have just had a thought about all the plagues that accrued. There is a natural way why each plague followed the last one.
The water turning into blood.

The experts say

That what coursed the river to turn red was that a volcano erupted. The sulphur and the darkness were perfect for breeding larva.
No wonder the frogs left the rivers it must have been like trying to swim through treacle. When the frogs left the river there food supply was no longer available that meant that their diet of no food killed them. The rotting carcases meant that the maggots eventually turned into biting insects. This was the course of the animal deaths and this was followed by boils. We then have the hail. This hail weighed up to one hundred weight. Along comes the locus which disappeared when the darkness came down. The last one that changed Pharaoh's mind was the death of the first born.

darkness why did this not happen when the volcano erupted, I mean to say it caused the sulphur phenomenon. The only explanation is what I thought had happened was that the wind changed direction.

Moses certainly was an incredible perSon that could predict ten plagues in the order that they came in. to our scientists it was easy to work how each one was the course of the next one. And he did this with ten plagues. How could any one, in that period fore caste these miracles.

The most asked question asked by the people, is if there is a God why is there So much suffering in the world. There were millions of people killed by the lord. There were the floods in the time of Noah. Then there was the raining down of fire and sulphur that destroyed the five cities, including Sodom and Gomorrah. Then we have the exodus by the Israelites from Egypt. This caused the death of the Egyptians by drowning. Why, why, why. Let us understand our lord is not of human form. So how can we expect him to have feelings? He created us as he created the whole universe. As he said in the beginning from dust to dust that is all we were. It was only when he sent his Son to Earth and he died on the cross that he forgive us because of Jesus' forgiveness of his famous words. Forgive them father for they no not what they have done. He could have made us with no feelings then we would have been zombies. It is because of feelings that we are what we are. If Jesus can forgive the wrongdoings of his executioners then surely we should be able to forgive those that trespass against us. This is the hardest thing to do, but in the end it would ease all are pain. By harbouring these feelings of hate we are destroying our selves. And if you believe in life after death then you must fore-give because you will be for-given. There is one thing that all the scientists agree with is that you cannot produce any thing from nothing no matter what it is. Now comes my problem if this is So what was it that created our LORD. With the billions of miracles that are happening every second why should I worry about that one?

SPACE AGE

What is space? And how old is the universe. I apologise but I am going to put forward a theory. Space has no past and there fore it is only as old as you see it. How have I come to this theory? Let me use my logic. If you have nothing then it does not exist. So it cannot get any older. The only way we can give space a time scale is since the universe came about. We can give the universe a time when it first happened, but the space between our sun's, planets, gases and satellites have not aged. Only all the objects in our universe have this ability to age.

Here's a thought. I don't know why I have not heard this mention, maybe I missed it. Let's have another look at the scenarios that have been put foreword. There is the one that we came from outer space. There is another theory that we originated from bacteria that crawled out of the sea. if there are any more that you know of you can put them into the equation. Let's go back to the beginning. And that applies to the both scenarios. Now here is the point. The dinosaurs got removed from the planet Earth and I would imagine every thing else on our beautiful planet ended up to the same fate. I mean I can't see that the dinosaurs were picked out and nothing else was. The biologists and the geologists said there was plenty of vegetation. Then there was the floods that covered the Earth. This apparently happened more than once. Now we have the ice age that the geologists have said that this happened more than once. I contradict this even though the geologists say that this is what happened. We Also have the destruction of the five cities of Egypt. This includes Sodom and Gomorra. My point if these things happened, and it has been found to be true, by the geologists. Now we come to the crunch. This must mean that the Earth's population, animal life, plants and bacteria would must start all over again from the beginning. Over and over again. There is only one way that this would be possible and that would must be a divine body as in a God.

What is there beyond our universe this is a long drawn out question. How about this, and this is not a theory. You might think this is a wild statement. Alright what I am basing this on is what the scientist, astrologers and physicists say. I would like to point out about black holes and there ability to hold stars, plants, and universes yet we can't see a speck. (They should have gone to spec save.) What is there in outer space that we can not see? There are two invisible facts and they are all around us on Earth. They are Kane's Force and sunlight. Materials that are in asteroids, planets and gases. Will not reach out there they are contained in our universe. We know that these materials travel at seventeen thousand mile per second and light travels at per second. That by my calculations means that matter will never outstretch lights speed. Working on that basic these two have travelled distance into outer space. I have not worked out how far this is but it must be when the first sun was created. We do not know when our Lord created the original sun So we must go by the time of the Big Bang. If we take that as fifteen billions years ago, multiplied by the speed of light, that's a long time ago. The sun's in the universe light up our planets, asteroids and gases. This of course is against all the atmospheres and elements. Surly if there was any thing in outer space it would be easier to detect.

The biologists have discovered the remains of prehistoric dinosaurs. Apparently they have worked out that the bones from a small dinosaur and giant dinosaurs are the same. Then they came up with that the large dinosaurs have evolved from the small dinosaur. To me its Sounds like Some one has it wrong.

The elephant they say have descended from the mammoths. With any species the offspring do not grow larger than there parents. There might be a small difference in a rare case but there young will revert back to there parents size. If this was not So then the elephant strain would have increased in size. We know that the human race has increased in height but that is because of our diet and health care. I don't think that the animal world have this capability. That to me means that species do not increase in size to their parents. This to me means that it is not possible for us to have evolved from bacteria. Let us look at the nearest things that we mustday that are like animals of the dinosaur's age. First the elephant. Now let us compare it with the nearest we can get to an animal from billions of years ago. The mammoth hello what is wrong the elephant is not bigger. I will give one more example. And this is more closely related than my elephant example.

The crocodile. Today's species is a lot smaller. Why do you think this is So?

The above statement concerning the dinosaurs was on the Quest channel on Monday 11/10/16.

I hope that Some of my theories will make the physicists see things in a different light.

Also Some of my logic way of thinking will put a different light on things.

Why do I think that gravity is a lot stronger than the sun's strength. as the scientists have said that sun's gradually lose there heat the further away you are from it. Now with gravity? The opposite happens. Let us look at Some thing here on Earth that can easily explain this. We will need the help of the greatest invention to mankind. The wheel if you spin a wheel then the further you are away from the hub the faster you go. And the bigger the wheel means you will go faster you don't must spin the wheel faster it is a well known fact,

Our galaxy/moon and the sun were created at the same time. They say that there are still signs of heat coming up from the moons interior. Now I have a question. The moon has its own gravity which is supposed to affect our tides. Now we know that Earth has a gravitational field that is stronger than the moons. There is no water on the moon but there is dust. So why is the dust not effected by the Earth's gravity. We have sand storms here in Britain that has come from the Sahara desert. Now if the moons gravitational pull creates our tides then why doesn't our gravity affect the moons dust as our gravity is four times stronger? My answer later. My real problem is they say that there is no atmosphere on the moon. Why not you have heat rising and it can't escape from the moons surface because of its gravity, what is going wrong? If you have hot air rising and the cold air descending then you should get clouds which would produce water. Everything is there to bring this to a conclusion So why doesn't it. The planets in our galaxy were formed by a sun exploding, those furthest from that blast were effected differently depending on their construction and the absolutely zero temperature they uncounted on their journey through space.

Comets and asteroids come from matter that has come from a sun exploding these, asteroids are held in position by the Kane Force. We now have a problem if this is So how it is possible to leave this force field and enter into the Earth's atmosphere. There can be only one of two logic explanation. The first one is we must take the astrologers point that they are colliding with each other. This then would make them into smaller fragments. This being So then they are lighter and therefore as gases that are lighter they would flung further out of there present position. This then would give them the ability to enter Earth's atmosphere. There is another scenario and this is mine. That is as the galaxies are expanding this would mean that they were nearer the circumference. And being lighter they would of course head away from there present position and Some would hit the Earth.

Gravity holds things into place and does not throw them out into other planets orbits to crash to the ground. So where have these comets and asteroids come from. There is only one explanation. They arrive in Earth's Solar system from a sun's explosion that occurred from the Big Bang. These comets/asteroids escaped from the nearest sun's before they had sufficient pull to hold them in their domain. As I have or will explain they are in the position that gases are in. they are travelling between the galaxies that means they are not effected by their sun's hold as other objects are. Also we need to understand that as they are from the Big Bang period it is possible that they were moving faster.

Our Solar system is one of the latest to be created. So let's take what would happen when our earth explodes. As we are the youngest galaxy in the universe yet we will be the first planet to explode. The reason for this is when our Lord created the earth it was in a later stage of its life cycle. We are millions of miles from our sun and we are travelling at seventeen thousand miles a second. Bang that's the end of our sun. Going by what we know about the universe expanding and what I have just said then the matter from our sun would be travelling at the same speed that we are So it would not affect us. What happens is another galaxy is formed So the matter would be controlled by their sun's. This would leave a void, or a black hole.

THE ASTEROID BELT

The scientists say that the asteroids that are entering our atmosphere have come from the asteroid belt. And Also as they arrive from all angles they have come from outer space. (This cannot be as my explanation on how galaxies work would not allow this.) As our Solar system is one of the latest Solar systems that have appeared in the universe all the matter must be getting smaller. To get to one of my points about asteroids is. We have a sun exploding and that matter is blasted in every direction. That means that these are heading for the black hole which means towards us if we are travelling towards the matter at seventeen thousand miles per second then the matter is travelling towards us at the same speed it will take half the time to reach us. And that is why they enter Earth from different angles. When the asteroids enter our atmosphere they do not head towards the Earth at twice the speed as in Newton's law on gravity the equal forces react against each other. How about this for a revelation? Why are these comets/asteroid not trapped in our atmosphere? There can be three reasons. One is their weight the second one is their speed and the last is their size.

Why would you, anybody or our Lord create another planet with life on it? Once you have created Something you don't want to do it again especially when you have done such a good job the first time. If you look at the way the earth with a different view on how things are going in a right mess you would not do it again.

I am now going to put forward a theory. The reason for this as there has never been one item printed discussed by any physicists, scientists, or any one at all interested in space, this is concerning space itself. When I first thought about this it seemed impossible. The only reason we think of space is because we have the universe. There is quite a bit of space. Sorry to go on a different track but I must edit it or I will forget. As the universe expands space does not. The space is just the same every second that means that if there was no universe there would be nothing. Think about it you would look up and there was nothing. There would be nothing to see as there would be no light to reflect any light. Now how about my theory, if there is nothing how can there be time. Did space start as a small nothing and then get bigger. Or has it all ways been there this is the most likely scenario. If you have nothing then you cannot expand it. This now gives us time in space there is no time. When you look up into space that is how old space is.

How about this for an eye opener? We have our universe. And the misrepresented view is that there is nothing there. I wish to contradict this view. What is there is Some thing that you cannot see. You might as well ask if you can't see it then it is not there. What are the two things here on Earth that we can't see but we know that it is there. It is of cause the sun's rays and gravity. As the universe expands is it then captured in the universes gravity? This Also covers the sun's rays. Now what is happening is that the sun's rays and gravity are already in space. But you can see no evidence as there is nothing there to pick it up. My explanation for this is (has the milky Ways centre got a Sun I think it has. (I believe that the physicists and the scientists believe it contains gases.) There is only one other explanation, and that is we must go back to time before the Big Bang. What did we have? There was of course only one thing and that is the original sun. If this sun was the original reason that

We had sun's rays and gravity then surly when the Big Bang occurred, surly this would cut off these two commodities. As the matter from the sun exploding was travelling at fifteen thousand miles per second and the sun's rays travelling at the speed of light that it does travel at then it is logic that the sun's heat would be long gone. That to my way of thinking is that there must be Some thing else supplying this need. As I have put forward there must be another sun to do this and the brightest thing in the sky is the centre of the Milky Way. There can be another explanation. And that is when the Big Bang accrued the centre of the sun was not blasted into space. Forget the gases we know that gases are hot but they do not produce rays of sunlight to the extent that a sun can produce. How can they give us the gravitational pull to keep our universe in tow? How's is this for a thought, is gravity stronger than the sun's heat. If you look at the universe you will see that it is still turning. But each galaxy/Solar system has its own heat force, and that is its own sun. This Also backs up my way of thinking that there is Something in the universe that is still working. Let us now take a look at gravity. Why do I think that gravity is stronger than the sun's heat? By the scientists, physicists and other bodies they say the sun's heat diminishes as it travels further into space I am going to agree with them. I bet that comes as a surprise, but do not bank on it. So were do I get that gravity is stronger. If you have a hub of anything as a sun's ability to produce this effect then if you consider a wheel spinning around the circumference spins faster So as you go further out into space then it would spin faster. That means that in our Solar system the planets further away from the sun are travelling faster than the planet Earth these are gas planets. This must Also apply to the universe. This Also explains why the gas giants are held in a global shape. There are lots of internet sites that try to explain gas giants. I have taken the Wikipedia site as this is the site that I usually use for any information. I have deleted any photos because they take up a lot of space. I hope my readers will be able to follow my replies and any contradictions I have about their theories on the subject.

A gas giant is a giant planet composed mainly of hydrogen and helium. Jupiter and Saturn are the gas giants of the Solar System. The term "gas giant" was originally synonymous with "giant planet", but in the 1990s it became known that Uranus and Neptune are really a distinct class of giant planet, being composed mainly of heavier volatile substances (which are referred to as "ices"). For this reason, Uranus and Neptune are now often classified in the separate category of ice giants. Jupiter and Saturn consist mostly of hydrogen and helium, with heavier elements making up between 3 and 13 percent of the mass. They are thought to consist of an outer layer of molecular hydrogen surrounding a layer of liquid metallic hydrogen, with probably a molten rocky core. The outermost portion of their hydrogen atmosphere is characterized by many layers of visible clouds that are mostly composed of water and ammonia. The layer of metallic hydrogen makes up the bulk of each planet, and is referred to as "metallic" because the very large pressure turns hydrogen into an electrical conductor.

The gas giants' cores are thought to consist of heavier elements at such high temperatures (20,000 K) and pressures that their properties are poorly understood. The defining differences between a very low-mass brown dwarf and a gas giant (estimated at about 13 Jupiter masses) are debated. One school of thought is based on formation; the other, on the physics of the interior. Part of the debate concerns whether "brown dwarfs" must, by definition, have experienced nuclear fusion at Some point in their history.

TERMINOLOGY

The term *gas giant* was coined in 1952 by the science fiction writer James Blish and was originally used to refer to all giant planets. It is, arguably, Something of a misnomer because throughout most of the volume of all giant planets, the pressure is So high that matter is not in gaseous form. Other than Solids in the core and the upper layers of the atmosphere, all matter is above the critical point, where there is no distinction between liquids and gases. The term has nevertheless caught on, because planetary scientists typically use "rock", "gas", and "ice" as shorthand's for classes of elements and compounds commonly found as planetary constituents, irrespective of what phase the matter may appear in. In the outer Solar System, hydrogen and helium are referred to as "gases"; water, methane, and ammonia as "ices"; and silicates and metals as "rock". Because Uranus and Neptune are primarily composed of, in this terminology, ices, not gas, they are increasingly referred to as ice giants and separated from the gas giants.

I am surprised to see in the above statements that they

ICE GIANT

Uranus, heavier nearer the sun my point. neptune furthest from sun

An **ice giant** is a giant planet composed mainly of elements heavier than hydrogen and helium, such as oxygen, carbon, nitrogen, and sulfur. There are two known ice giants in the Solar System, Uranus and Neptune.

In astrophysics and planetary science the term "ices" refers to volatile chemical compounds with freezing points above about 100 K, such as water, ammonia, or methane, with freezing points of 273 K, 195 K, and 91 K, respectively (see Volatiles). In the 1990s, it was realized that Uranus and Neptune are a distinct class of giant planet, separate from the other giant planets, Jupiter and Saturn. They have become known as *ice giants*. Their constituent compounds were Solids when they were primarily incorporated into the planets during their formation, either directly in the form of ices or trapped in water ice. Today, very little of the water in Uranus and Neptune remain in the form of ice. Instead, H_2O primarily exists as supercritical fluid at the temperatures and pressures within them.

Ice giants consist of only about 20% hydrogen and helium in mass, as opposed to the Solar System's gas giants, Jupiter and Saturn, which are both more than 90% hydrogen and helium in mass.

The Milky Way's sphere is the centre of the universe. Andromeda galaxy is part of the Milky Way.

Are the physicists, scientists, and astrologers, trying to pull the wool over our eyes concerning their theories over the universe? They are supposed to be our teachers. How can we possibly take them serious? and if we are not learning any thing new how can they be when they are repeating what they have been told is there not one of them that can think fore there selves. it looks like they are leaving it to me, god help us. Maybe they are worrying about their jobs as are all NASA employees. Surly with all their brains they could go into innovations that we could benefit all of mankind.

They say evolution takes take's billions of years. The reason that living thing evolve is because of their environment. If their environment does not change they have no reason to evolve. I mean what would be the point. If there is sudden change in their environment they must adjust in a matter of days or otherwise they become extinct. It only takes one male and one female to survive and the species goes on. You might think this is ridicules So read on.

There are Some islands that have materialised only recently, this is from volcanic eruptions. And they were habited by marsupials there was none of their natural food on this island. If they had to rely on evolution they would have all died. So within days they had to adjust. They swam

Bang Goes That Theory

under water to reach plankton. This then became their basic food. I can imagine that in another hundred years' time Someone will go this island that is now covered in vegetation will discover these marsupials and they will say that they are entirely different from their cousins from miles away and have evolved over hundred or even billions of years.

How's this for a revelation. I have noticed that Something that will put beyond doubt whether there is a creator not, and you will realise when I give you the facts. These are theories the physicists, scientists and geologists and all the other Boffins have come up with. I know I am not an expert on the universe but I have taken it in what they have been saying. I have left this tremendous fact to the last because I have only just thought about it. I think this covers more in the line for geologists. Before Noah's time the human race had the abilities to build structures that honoured our LORD they reached into the heavens. Man's structures displeased our LORD as the people were looking at these churches putting man's achievements as the same as our LORD'S. God was displeased and split the land into many parts. Not only did he do this but he gave the people a different language So that they could not communicate with each other. Some of these people/ tribes were on the same land, this meant they could not communicate with each other and disharmony and wars occurred between the different tribes. Now we come to my observation, but first we will look at what the geologists say. The land was all one mass and over time the rivers and seas corroded the land mass and they became as they are today. As we can see this looks like a feasible explanation to come too? The bible says the parting of the land mass happened before Noah time. Looking at the way nature works and the time span to me this seems reasonable. Now let us look at the scientists and geologists way of looking at things. There explanation of there being one mass and over time the seas and rivers and I should imagine the wind would Also have an effect on the land I had forgot about the wind but I must put it in to give this explanation every chance of succeeding. Now we have the time scale. We must go back to the forming of the Earth. This by the geologists reckoning is round about fifteen billion years ago. That's a lot of corrosion over a long time. I know the experts say that the weather then was a lot fiercer than it is now So there should be a lot of corrosion quicker. And yet Britain is losing a lot of coastline in the present day. Let us now look at this over billions of years.

There is a lot of land that must go into the sea. Now if this is the case I do not think that the beaches and the sea beds around our coast line should look the way it is. There should be hardly any rocks only where there have been signs of earthquakes. We can see by all the sand on the beaches and on the deserts what happens over time. There is another point, and this is what has made up my mind and I hope that the scientists and the geologists will see the logic of my explanations and how that the only way that the land could have parted is by a creator

The Earth has a circumference of twenty-two thousand miles this gives us a diameter of fourteen thousand miles. If the land mass was all in one and it was corroded away over time I would like the experts to tell me the following mathematic equation. The Earth surface consists of seventeen five percent water and twenty five percent land mass. Has anyone looked at the worlds map recently I don't think there has been much change to it. As can be seen all the continents when pushed together are interlocking. This has happened by the sea eroding the coast lines. In the beginning the weather then, to put it mildly would make our global warming look like a walk in the park. This has taken fifteen billions years to do this. In that time volcanoes, earth quakes and the moving of the Earth's plates must have made quite a change to the coast lines. Do you honestly think that they would an interlocking land mass, of course not? We can see the affect that global warming is doing now, and that is in the last five years. There must be times in the last billion years that there were storms that make our global warming look like one drop of rain in a bad winter.

You will notice that there is a lot of sea between the different continents. Have you got your thinking caps on? If the land mass was all one, then all the water between these continents would of cause must be filled in with land. This would mean that the amount of area that is left must be filled in with water becomes less. I am no mathematician but that means that where this area is available then it must be deeper to take this amount of water. Now we have a problem with the water. We must find a place for all this water. Maybe if we get Some spades and start digging we can make this channel deeper. You might think this is a stupid idea, it is. The only other explanation is that the seas are deep enough to take this water. Wrong owing to the depth that would be needed they would be that deep they would reach into the plasma. I have got that wrong it would reach the core. The following 24,901sum needs working out give me a minute. I am not bragging only joking. The earth's circumference is 22,000 miles. =8.300miles radius depth. Water depth 10,994 If you can recall that I would explain why I pointed out that the crust of the earth was very thin at a certain point.

You might think that evolution is good thing. I know by experience that this is not So. You might as well say I have run out of grey cells. I speak from my bitter experience. You know my opinion on what the experts have come up with that we have descended from the apes. I didn't. But just to make my point we will consider this. Now I hope I have really got you going, and you are now burning up Some grey cells. What can it be? I really am enjoying this. Has this go Something to do with the length of their arms. You might think I am missing swinging in the trees. No try again. You have part of it right, it is the arms. When you get to my age you will realise why this is disadvantage. I get really annoyed, and as time goes by it gets to the point that I now need help. Do not get me wrong this Also the blight of the young. It is putting on your Socks and shoes. How easy it would be with longer arms. And think of the advantage getting objects off the top shelf.

We could look at it in different way, how it would be if we had shorter legs that could Solve the problem of the shoe scenario, but not the higher shelf problem.

Ocean - Wikipedia
https://en.wikipedia.org/wiki/Ocean
Its average depth is about 3,688 meters (12,100 ft), and its maximum depth is 10,994 meters (6.831 mi) at the Mariana Trench. Nearly half of the world's marine waters are over 3,000 meters (9,800 ft) deep. The lowest depth is said to be two miles deep. The vast expanses of Deep Ocean (anything below 200 meters or 660 feet) cover about 66% of Earth's surface.

With all the different religions were their customs discriminates against any other perSon, for any reason, whether it is on religious grounds, gender or race has nothing to do with the law of God. These customs are made by man. And in all cases, you will find that they are made by man to dominate any one in their Society that are not strong enough to defend their selves. To me this means that they are bullies. There is great condemnation against domestic violence then this should Also apply to these traditions. It is time that these traditions were banned, as they go against the English law, of the individual's right. Why you might ask. Let us take the Muslim tradition. The women are told what to wear. Knowing woman, do you not think they are any different than the western females, maybe I have it wrong I don't know. What I am trying to say is everybody tries to look good whether they are females or males. There is Also the Jewish tradition of circumcising the boys. This is barbaric and should not be allowed in Great Britain, or anywhere else on planet Earth. I just thought I would give my opinion. Another note on the subject. *E*ven the lowest form of life on the planet tries to attract the opposite sex and that is by preening itself or showing of in one manner or another. All traditions of all races should not be allowed when they are there to discriminate against any perSon by dominating them, whether by gender, sex preference or religion.

The contradictions of the below will be at the end of these predictions. I will reply to these statements given and point out why I have come my decisions based on earlier information given by scientists, physitrists and astrologers.

A NASA conception of the collision using computer-generated imagery.
Here we go again using computer generated images that confuses us that do not have much knowledge on space that we believe what these images are what is happening.

The Andromeda–Milky Way collision is a galactic collision predicted to occur in about 4 billion years between the two largest galaxies in the Local Group—the Milky Way (which contains the Solar System and Earth) and the Andromeda Galaxy The stars involved are sufficiently far apart that it is improbable that any of them will individually collide. Some stars will be ejected from the resulting galaxy, nicknamed *Milkomeda* or *Milkdromeda*.

While the Andromeda Galaxy contains about 1 trillion (10^{12}) stars and the Milky Way contains about 300 billion (3×10^{11}), the chance of even two stars colliding is negligible because of the huge distances between the stars. For example, the nearest star to the Sun is Proxima Centauri, about 4.2 light-years (4.0×10^{13} km; 2.5×10^{13} mi) or 30 million (3×10^7) Solar diameters away. If the Sun were a ping-pong ball, Proxima Centauri would be a pea about 1,100 km (680 mi) away, and the Milky Way would be about 30 million km (19 million mi) wide. Although stars are more common near the centres of each galaxy, the average distance between stars is still 160 billion (1.6×10^{11}) km (100 billion mi). That is analogous to one ping-pong ball every 3.2 km (2.0 mi). Thus, it is extremely unlikely that any two stars from the merging galaxies would collide.[5]

Black hole collision

The Milky Way and Andromeda galaxies each contain a central supermassive black hole (SMBH), these being Sagittarius A* (ca. 3.6×10^6 $M_\odot$) and an object within the P2 concentration of Andromeda's nucleus (1–2×10^8 $M_\odot$). These black holes will converge near the center of the newly formed galaxy over a period that may take millions of years, due to a process known as dynamical friction: As the SMBHs move relative to the surrounding cloud of much less massive stars, gravitational interactions lead to a net transfer of orbital energy from the SMBHs to the stars, causing the stars to be "slingshot" into higher-radius orbits, and the SMBHs to "sink toward the galactic core." When the SMBHs come within one light year of one another, they will begin to strongly emit gravitational waves that will radiate further orbital energy until they merge completely. Gas taken up by the combined black hole could create a luminous quasar or an active galactic nucleus. As of 2006, simulations indicated that the Sun might be brought near the center of the combined galaxy, potentially coming near one of the black holes before being ejected entirely out of the galaxy.

Based on data from the Hubble Space Telescope, Milky Way galaxy and Andromeda galaxy are predicted to distort each other with tidal pull in 3.75 billion years. The Andromeda Galaxy is approaching the Milky Way at about 110 kilometres per second (68 mi/s) as indicated by blueshift. However, the lateral velocity is very difficult to measure with a precision to draw reasonable conclusions: a lateral speed of only 7.7 km/s would mean that the Andromeda Galaxy is moving toward a point 177,800 light-years to the side of the Milky Way ((7.7 km/s) / (110 km/s) × (2 540 000 ly)), and such a speed over an eight-year time frame amounts to only 1/3000 of a Hubble Space Telescope pixel (Hubble›s reSolution≈0.05 arcsec: (7.7 km/s)/(300 000 km/s)×(8 y)/(2 540 000 ly)×180°/π×3600 = 0.000 017 arcsec). Until 2012, it was not known whether the possible collision was definitely going to happen or not.[7] In 2012, researchers concluded that the collision is sure using Hubble to track the motion of stars in Andromeda between 2002 and 2010 with sub-pixel accuracy. Andromeda's tangential or sideways velocity with respect to the Milky Way was found to be much smaller than the speed of approach and therefore it is expected that it will directly collide with the Milky Way in around four billion years.

Such collisions are relatively common. Andromeda, for example, is believed to have collided with at least one other galaxy in the past,[8] and several dwarf galaxies such as Sgr dSph are currently colliding with the Milky Way and being merged into it.

These studies Also suggest that M33, the Triangulum Galaxy—the third largest and brightest galaxy of the Local Group—will participate in this event too. Its most likely fate is to end up orbiting the merger remnant of the Milky Way and Andromeda galaxies and finally to merge with it in an even further future, but a collision with the Milky Way before it collides with the Andromeda Galaxy or being ejected from the Local Group cannot be ruled out

Two scientists with the Harvard–SmithSonian Center for Astrophysics stated that when, and even whether, the two galaxies collide will depend on Andromeda's transverse velocity. Based on current calculations they predict a 50% chance that in a merged galaxy, the Solar System will be swept out three times farther from the galactic core than its current distance. They Also predict a 12% chance that the Solar System will be ejected from the new galaxy Sometime during the collision.[9] [10] Such an event would have no adverse effect on the system and the chances of any Sort of disturbance to the Sun or planets themselves may be remote.

Excluding planetary engineering, by the time the two galaxies collide the surface of the Earth will have already become far too hot for liquid water to exist, ending all terrestrial life; that is currently estimated to occur in about 3.75 billion years due to gradually increasing luminosity of the Sun (it will have risen by 35–40% above the current luminosity).

Possible triggered stellar events[edit]

When two spiral galaxies collide, the hydrogen present on their disks is compressed producing strong star formation as can be seen on interacting systems like the Antennae Galaxies. In the case of the Andromeda–Milky Way collision, it is believed that there will be little gas remaining in the disks of both galaxies, So the mentioned starburst will be relatively weak, though it still may be enough to form a quasar.

Merger remnant[edit]

The galaxy product of the collision has been nicknamed *Milkomeda* or *Milkdromeda*. According to simulations, this object will look like a giant elliptical galaxy, but with a centre showing less stellar density than current elliptical galaxies. It is however possible the resulting object will be a large disk galaxy, depending on the amount of remaining gas in the Milky Way and Andromeda.

In the far future, roughly 150 billion years from now, the remaining galaxies of the Local Group will coalesce into this object, that being the next evolutionary stage of the local group of galaxies.

Theoretical studies indicate that most supernovae are triggered by one of two basic mechanisms: the sudden re-ignition of nuclear fusion in a degenerate star or the sudden gravitational collapse of a massive star›s core. In the first instance, a degenerate white dwarf may accumulate sufficient material from a binary companion, either through accretion or via a merger, to raise its core temperature enough to trigger runaway nuclear fusion, completely disrupting the star. In the second case, the core of a massive star may undergo sudden gravitational collapse, releasing gravitational potential energy as a supernova. While Some observed supernovae are more complex than these two simplified theories, the astrophysical collapse mechanics have been established and accepted by most astronomers for Some time.

If you have noticed the first word in the paragraph says, theoretical. Now I will blow all of this into its true prospective.

There was a program on the B.B.C. 06/05/17. Walking through Time. Britain's last mammoths. These bones where found in the north. The bones of these mammoths are said to be 12,000 years old and they roamed there at that time. The remains in the rocks showed that there was a food supply for these animals, Also their stomachs contained food So they did not die of starvation. They have worked it out that an ice age was probably the course of their demise. The reason for their working this out was because of the way the Earth's plates have moved, and the movement of the plates in an eastern direction. So what I deduct from this is if you think what they have just said that there was an ice age fifteen thousand years ago. The most outstanding thing is there statement that the reason these mammoths have decided to move north was of the earth's plates. In fifteen thousand years how far have these plates moved? To my reckoning it is very slow. I am no mathematician, but I reckon about a foot and that is being generous. Maybe I am wrong I will let you work it out.

WRONG.

You are probably asking yourself how I have come to this conclusion. Let us go back and do Some mathematics. Twelve thousand years ago the mammoths were grazing peacefully in their habitat. They did not realize what they were doing. But they were going to move north over the next twelve thousand years. When I say they moved they did not travel by foot. I hope I have this right. These bones that they found were fifteen thousand years old when they died it got a bit to cold for them. So they died fifteen thousand years ago. Bingo. They were not in the north at that time they were fifteen thousand miles away when they died they were not alive in the north because they were not there. There is another way to look at it fifteen thousand years ago they lived in the north and then there was an ice age.

There are three main movements of the Earth's crust and to go into these would take up quite a bit of time for my readers I hope this book will be read by more of the public and for those that need more of a scientist view can look this up or learn from their academic studies. The maps below show these three movements. And my small amount of knowledge of this subject plus mathematically this is not possible. I will try to keep it at its basic. 12,000 years ago these huge mammoths were roaming on the earth's surface. Then they died from a very serious case of frost bite. An ice age does not happen in twelve thousand years. The experts say that an ice age takes a long time to effect the enviroment. I take this as million of years. Let us now look at todays discovery of these mammoths. They might be twelve thousand years old but that was twelve thousand years ago. The reason that they have arrives in the north is because of the earth's plates and how they move. The Southern plates were pushed over the nothern plates. As it is only twelve thousand years ago then there must be a fault in the earth's crust in the near proxsimity. There are lots of minor fauls but these are uually caused by the earth sinking at certain points and the surrouding surface not moving. This is usually caused by a part of the rocks being of a Softer consistancy and therefor being washed away. These are entirely different from the movement of a major fault. By my way of looking at things this was then a major fault that occurred over billions of years ago and in that billion of years it finally arrived in the north. Twelve thousand years ago these mammoths died.there was a baby mammoth that had food in its stomach. So it didn't die of starvation. Maybe the digestive system was not working as it should and it died of a bad case of constapation. Was man the course of it's demise. If we look at todays animal kingdom maybe we can learn Something. What courses spieces to become extinct. Man does it by destroying there enviroment. it must be a freak in nature it would not take long for a quick change in the enviroment. a short period of a drought or a short ice age. I think a ice age would only be a matter of months not years. the youngest would die first because of lack of milk from their mother. and then the next ones to die would be the the previous litter and they would till have traces of food in their stomachs.

There have been at least five major ice ages in the earth's past (the Huronian, Cryogenian, Andean-Saharan, Karoo Ice Age and the Quaternary glaciation). Outside these ages, the Earth seems to have been ice-free even in high latitudes.
Ice age - Wikipedia
https://en.wikipedia.org/wiki/Ice_age
this works out at three milion years between ice ages I am getting good at this dividing. Lets take it that an ice age was just around the corner you don't need to go and have a look I am speaking methatholicly.
Transform boundaries

TranSorm. Boundaries (Conservative) occur where two lithospheric plates slide, or perhaps more accurately, grind past each other along transform faults, where plates are neither created nor destroyed. The relative motion of the two plates is either sinisterly (left side toward the observer) or dextral (right side toward the observer). Transform faults occur across a spreading centre. Strong earthquakes can occur along a fault. The San Andreas Fault in California is an example of a transform boundary exhibiting dextral motion

Divergent boundaries

Divergent boundaries (Constructive) occur where two plates slide apart from each other. At zones of ocean-to-ocean rifting, divergent boundaries form by seafloor spreading, allowing for the formation of new ocean basin. As the ocean plate splits, the ridge forms at the spreading centre, the ocean basin expands, and finally, the plate area increases causing many small volcanoes and/or shallow earthquakes. At zones of continent-to-continent rifting, divergent boundaries may cause new ocean basin to form as the continent splits, spreads, the central rift collapses, and ocean fills the basin. Active zones of Mid-ocean ridges (e.g., Mid-Atlantic Ridge and East Pacific Rise), and continent-to-continent rifting (such as Africa's East African Rift and Valley, Red Sea) are examples of divergent boundaries.

Convc Convergent boundaries (Destructive) onvergent bounderiesergent bouderies

Convergent boundaries (Destructive) (or active margins) occur where two plates slide toward each other to form either a subduction zone (one plate moving underneath the other) or a continental collision. At zones of ocean-to-continent subduction (e.g. the Andes mountain range in South America, and the Cascade Mountains in Western United States), the dense oceanic lithosphere plunges beneath the less dense continent. Earthquakes trace the path of the downward-moving plate as it descends into asthenosphere, a trench forms, and as the sub ducted plate is heated it releases volatiles, mostly water from hydrous minerals, into the surrounding mantle. The addition of water lowers the melting point of the mantle material above the sub ducting slab, causing it to melt. The magma that results typically leads to volcanism. At zones of ocean-to-ocean subduction (e.g. Aleutian Islands, Mariana Islands, and the Japanese island arc), older, cooler, denser crust slips beneath less dense crust. This causes earthquakes and a deep trench to form in an arc shape. The upper mantle of the sub ducted plate then heats up and magma rises to form curving chains of volcanic islands. Deep marine trenches are typically asSociated with subduction zones, and the basins that develop along the active boundary are often called "foreland basins". Closure of ocean basins can occur at continent-to-continent boundaries (e.g., Himalayas and Alps): collision between masses of granitic continental lithosphere; neither mass is sub ducted; plate edges are compressed, folded, uplifted.

as can be seen by the three different ways that the plates move and the differrant ways the faults react to these pressures.

Magnetic field the magnetosphere is the region above the ionosphere that space. It extends several tens of thousands of kilometers into space, protecting the Earth from the charged particles of the Solar wind and cosmic rays that would otherwise strip away the upper atmosphere, including the ozone layer that protects the Earth from harmful ultraviolet radiation.

This is my explanation where and how magnet fields came about the original sun contained all the elements and metals and these were spread over the universe at the time of the Big Bang. We know what makes a metal magnetised. This is a metal that has an electric charge passed through it. The sun has an electric field this than is the charge that converts the metals into a magnet. Bang that's the sun exploding, and all the planets and sun's contain a magnetic field.

They Also put forward that the magnetic fields protect our ozone layer by travelling into space. Sorry this is not possible. You know about Isaac Newton's law concerning equal forces. I have pointed out about Isaac Newton's law. This law means that nothing leaves the earth's atmosphere otherwise there would be none left unless it can reproduce the lost product. If there was Solar wind rushing towards the Earth at 400 km/s, and 700 km/s for the fast wind why did this not affect the

The theories that will be put forward by the powers that give us all the information on space are only theories. As I have explained about the gas giants and gases that are in our universe how can the Solar winds can travel willy nilly everywhere. There must be Something that controls them as everything else in the universe is controlled Some way. The northern lights are affected the Solar winds that is why they react in the way they do or So the experts say. By what I understand about the Solar winds is that they are magnetically charged as in positive and negative.

The record holder is Also easy to find, it's the New Horizons mission to Pluto and the Kuiper belt.

16.26 kilometers a second (that›s about 36,000 miles per hour), --

--plus, a velocity component from Earth's orbital motion (which is 30 km/s tangential to the orbital path).

Why has nobody explained why it is with billions of sun's in all these galaxies they do not light up anything outside their galaxies? This is because of Isaac Newton's theory on gravity. He didn't realise and nobody else has realised what is happening to the sun's rays. They are governed by the same reason that gravity is governed, and this is equal forces cancel each other out. If you shine two lights onto a surface the light will be twice as bright, but if you shine two lights at each they will not be twice as bright. This is what happens with the sun's rays in each galaxy reaches the point when it meets the next galaxy. I mean if it didn't then the whole universe would be one blinding light and we not be able to see anything. How's that for a revelation. This Also applies to magnetic forces and any other force that comes from a sun's power. If this did not occur than the universe would cease to exist, in fact it would never have started.

There were two Russian scientists that lived in a flat in America. They had a more confiscated computer system then the American defence, or the C.I.A. department system. They spent twenty-five years trying to work out a formula that would work out a mathematical Solution to PI. They couldn't do it. The nearest all the mathematicians have come up with is 22 over 7 for every negative there is a positive. The mathematicians are always saying that everything can be worked out mathematically. I have looked at this for about one minute, Sorry about the time scale, *I mean one minute!* It was more like thirty seconds and came up with my mathematical Solution. That is the diameter of a circle is seven inch. The circumference is twenty-two inch. By using the mathematic system that is being used means it cannot be divided. This Also appliers to ten divided be three. This is not a true. They are all looking at it in the wrong context. As with all professions, and in any other kind of work place, they are taught their job/profession, and that is what they do. I will reveal the mathematical way it should be done. I have taken notice of what all the coaches in all the sports sections say and that is when things are going wrong go back to the basics. So to keep you on your toes this will be reveal at a later date. Sorry to keep you waiting but I have a lot on my plate with this book, and I need to do a lot of studying. ha that's a joke. How was I able to work out in a few seconds what the top mathematicians and physicists have not been able to do in since maths became a tool for mankind.

The mathematicians have managed to work out how to convert Celsius to Fahrenheit. They did not take a straight path but went about it in from a different direction.

P.I.

There are three different equations that must be used. But before we go into that what we will do is give a child of nine a cake that has a radius of seven inches and the circumference is twenty-two inches. Ask the child to cut it in three pieces. Now they will not be three equal sizes. If they had the use of a rule or a computer I wonder what the outcome would be. But you can get my drift. Now let us do it in a more mathematical way. We can do it two ways. One is we can use the cake as one example. The circumference is twenty-two inches. Divide by three = three and one over. Now here's the trick you divide the one which is of cause tenths. You might say the idiot we are back to square one. True but what you now do is change the measurements into ninths, bingo. To make thinks easer why don't we make the twenty-two into ninths? And you could divide them by ninths. That gives us the 198 ninths dived by nine =12 ninths. To make the sum easier, divide 10 by 3. Change the 10inch into thirds = 30 thirds divide by 3 =10 thirds.

The geologists have unearthed the bones of our ancient ancestors. The previous bones dated back 200 thousand years ago. These they new findings they say go back 300 thousand years ago. What does this tell you about evolution? I hope you have been taking notice about my revelations about evolution. I will give you a quick reminder. it concerns the beetles in a zoo and the immediately change they took in there billion years of life on this planet. They went from a dung diet to a fruit diet instantly. And as I have said the insect and animal families when the environment changes quickly they must adjust to another diet, move with the changing seasons or become extinct. They do not have the ability to go from one way of life to another way that is far more advanced. We know that it takes million of years to evolve. This 100, thousand years difference in the extra time that man has been on this planet. What this tells you gives us one more reason why there is a creator. Just think about it. 300, thousand years ago man was on this planet. There are two ways of looking at this. If man has evolved from the ape So how long did this take? Going by how long evolution takes it must be billion of years. Okay we now have man. The first thing he must learn is not to be afraid of fire and to use it to his advantage. Also he loses all his hair, So he needs to clothe himself. And he must get down from the dam tree. Okay let's get Some order to this. First, he gets out of the tree. This is a godsend to the ape's predators. These most likely would be lions. he manages to do this by one means or another. He must then find a new place to live. We know that the first man lived in caves. There was volcanoes So this is one way he learned that the fire was warm. He still needed pyjamas of a night and sports wear to go hunting. Did he go hunting for any apes that were left swinging in the trees? If they all started running about there would not be any swinging apes left. As there are still apes around us we have an idea that Some backward apes continued their tree activities. This took billion of years then suddenly wham man gets a brain. Adam was supposed to be the first man to have this ability. The way man has advanced since the time of Adam. Remember the creation of Adam was at in one instance. Before this, man was as thick as a brick, or two short planks. When Adam was banned from the Garden of Eden he had to supply all his own food and as we know clothes. Since this time man had advanced more in the last three to five to seven thousand years than he has in the previous billion of years. This is not logic. We will go back the time scale has gone back another 100 thousand years that man has trod on the Earth and with all this extra time his intelligence had not moved. So much for apes evolving into humans. I hope you get my point that there had to be a creator.

Andromeda Galaxy was formed roughly 10 billion years ago from the collision and subsequent merger of smaller protogalaxies. This violent collision formed most of the galaxy's (metal-rich) galactic halo and extended disk. During this epoch, star formation would have been very high, to the point of becoming a luminous infrared galaxy for roughly 100 million years. Andromeda and the Triangulum Galaxy had a very close passage 2–4 billion years ago. This event produced high levels of star formation across the Andromeda Galaxy's disk – even Some globular clusters – and disturbed M33's outer disk.

Over the past 2 billion years, star formation throughout Andromeda's disk is thought to have decreased to the point of near-inactivity. There have been interactions with satellite galaxies like M32, M110, or others that have already been abSorbed by Andromeda Galaxy. These interactions have formed structures like Andromeda's Giant Stellar Stream. A galactic merger roughly 100 million years ago is believed to be responsible for a counter-rotating disk of gas found in the centre of Andromeda as well as the presence there of a relatively young (100 million years old) stellar population.

As can be seen in the above statement it says that the Andromeda constellation is ten billion years old, and it is the oldest galaxy in the universe. This galaxy contain one trillion (1012) stars:[11] at least twice the number of stars in the Milky Way, which is estimated to be 200–400 billion.[14] 2006 observations by the Spitzer Space Telescope revealed that Andromeda contains The mass of the Andromeda Galaxy is estimated to be 1.5×10^{12} Solar masses, while the Milky Way is estimated to be 8.5×10^{11} Solar masses.

The Milky Way and Andromeda galaxies are expected to collide in 4.5 billion years, eventually merging to form a giant elliptical galaxy'or perhaps a large disc galaxy. The apparent magnitude of the Andromeda Galaxy, at 3.4, is among the brightest of the Messier objects, making it visible to the naked eye on moonless nights,[18] even when viewed from areas with moderate light pollution.

REPLY

I do not know if my division is going by the wayside but looking at the above facts my sum totals seem to be wrong. Let me point out what they have said about the two different galaxies the age and the sizes. Working out their ages we have just been told that the Andromeda galaxy is the oldest at ten billion years. I believe that the Milky Way was the start of our universe. This was after the Big Bang. They say the universe is fifteen billion years old. Have you worked it out? - ten billion years from the above fifteen billion years. I will must get my calculator out as maths is not my best subject. Here I go again trying to fool my readers that maths is the only subject that I am rubbish at. I get a difference of five billion years. That is one part of the scenario Solved. The next problem is the number of stars in each of the galaxies. As I have pointed out earlier the size of matter that was produced from the Big Bang as in the Andromeda galaxy had the largest of this matter and the sun's took longer to hold this matter into its galaxy. How these experts have come to the fact that it is older beats me. I mean they keep finding fault in lots of previous finding.

RAINBOWS END

Do not think you will find the end of the rainbow here; the heading is my answer to the statements by the physitrists and scientists. They are So precise on the angle being 42 degrees we will take that as a fact. Where we are standing we look up and see a rainbow at 42 degrees. The sun is behind me and it is rays are hitting the raindrop at 42 degrees. Each raindrop has no connection to its neighbour. Remember the angles we are working with are in the deudecillions. What we need to do is to eradicate the distances. This means that the reflection from the next raindrop is further back. The angle will not be 42 degrees. It will be at a larger degree and this will be the case for the next raindrop and So on until you come to the raindrop that is at the same level. This means that the rainbow is seen at the beginning of the rainfall

It has not only the 42 degrees angle we must think about we must put the distance that the rainfall is away. We can have the sun behind us and there are many times there is no rainbow. The sunlight rays must be hitting these raindrops Somewhere at forty-two degrees. So why do they only show up when they are far away. We know you can have them nearer as seen from a waterfall or a water spray from any Source. These rainbows will be at a different size. This scenario is working against logic the nearer an image is then it should be bigger. There is another thing that the physiatrists must work out for me. We need the sun behind us. We Also need a rainfall. Now we have these two scenarios and we can see a rainbow.

Let us move two hundred metres nearer. *The sun's rays will still be capable of hitting the rainfall at forty-two degrees. It*

We can see he colours around the sun and these are all the way around, but we cannot see the ones that are directly in the centre of the sun. We must use the physics forty-two degrees angle. If this is correct there is Something wrong, there is no sunlight behind us So how can we see a rainbow. Well children Sorry readers if you have been taking notice you will realise that I have stated that these are gases and the brightness of the sun's rays obscures th colours. Why do we only see the colours when a light is shone at it or as in the case of Einstein seeing the colours when the light shone through a hole? The same principle applies to both scenarios. It is because the rays of the sun's white rays are reduced as in deflected or more than likely the white light is reduced thus allowing the colours to be seen. The prism does not show any white light only the colours this is done because the white light is deflected as can be seen in the prism the white light will be deflected at the same angle that it hits the prism. You will must find a background that shows the sun's white rays. Why are the colours that are shown in a prism not shown as a true deflexion as in the angle leaving should be the same as it enters. Here is a smack in the eye for scientists, astrologers, physitrists and all those who think they have Sorted out the reason that we can see a rainbow the colours of the rainbow can be seen at forty-two degrees. That means that all the colours are at forty-two degrees. Ha, I have a problem with this. The forty-two ⁰ has magical powers. Why you might ask. The raindrops must contain all the colours So how they can define which colour is in which raindrop So that all the colours appear in a bow.

How much water is in the ocean? I believe its 97% of the earth's crust it is estimated that it covers 70% of the land area.

The universe as can be seen from all the photos taken of the universe they are all taken from the same angle. This does not give a true picture. You would need a 3d photo for this. So now we must use a bit of common sense to see what it would look like from this angle. We need to turn the photo we have now. Let us look at it from the Earth's position. We twist the photo upwards as far as it can go. Now we have the Earth at the top of the photo. This means that as the Earth is a second-generation galaxy it should appear above the other galaxies. As can be seen the Earth is in the Southern regions of the universe. There are Some lone galaxies that are not in the regions of the Milky Way or the Andromeda region. As in any explosion there are particles that will travel a lot further than others.

Our nearest galaxy is the Provima B. the Proxima Centrian is the name given to its sun? There's a problem. Their sun is smaller than the Earth's sun. This is not possible as the Earth's are the second generation and was created after all the other sun's then it should be smaller. We could say after the Big Bang all the different sizes of matter that was blasted into space you would have all different weights. would not be affected by weight as they were travelled were there was no gravity. This is true but as they travelled through space the sun's Penny Force would grab hold of them and the lightest would be the first ones that they would grab hold of. This is physics or a scientific fact. Let us clear up after ourselves and put everything back to normal. Billion of year's later we will see that the universe has disappeared and there is only one small galaxy. Does this mean that the universe has been gobbled up by giant galaxies or black holes have pulled all this matter into its void? If I have your attention then I will tell you later what has happened. That was mean wasn't it? Have you waited long enough then I will tell you the answer. The universe is expanding. You would think that this would show that over billions of years the universe should be a lot larger than is, yet all the galaxies are clustered together. Is this an imploding scenario as the physitrists said, or is it black holes on the look out for food and found it? If this is the case, then we can expect this course to carry on in the same format. And puff there is nothing left. This is what I predicted but my reason for this is not the same as the physitrists. I will give you the reason why I have come to my conclusion. If you are standing at a main train station and there were lots of trains pulling in. We are looking at these train lines from a long distance away. The lines were converging together. Does this mean that the station has gobbling them up like a gas planet or a black hole, of course not? This is what the astrologers are trying to make you believe. We know these are only theories but if you are being taught about the workings of space, why can they not use a bit of logic. This is no way for students to be taught. We all know about the train line scenario, well most of us do. Why haven't the boffins that study space not realised that this is same thing that is happening to the galaxies? I hope that Somewhere in the bowels of the space exploration they will use Some of my logics. If students or pupils must be taught let us give them Some correct knowledge and not fill their brains with unconfirmed theories. As I have brought to your notice light reacts the same way that gravity does. As at the present time it appears that there is lighter coming from the Milky Way. This is because of the train line scenario. As each galaxy goes behind or next to the others, the light does not increase nor does its glare diminish. In time it will be that far away and as the astrologers have said the sun's heat and therefore its light will be diminished. We will not be able to see anything. This does not register with me

I hope you can understand my reasoning for the above statement. The first sun had the ability to create Penny Force, and heat from its rays. So what we have now is these two elements in space. Please note these elements are in space there is nothing else. These two elements are travelling at the same speed that is in an outward direction. They are of course the speed of light and the force of the Penny Force. This tells me that in time the Penny Force will of course become faster/stronger than the speed of light.

We will now see why there is no reason to fear asteroids or comets being the size that could destroy Earth. The American space ship that went to Mars took pictures of the planets surface. This surface was the replica of the Earth's surface. There where places that just looked like there had been a river flowing there and then there was a drought. This left the river bed with just the rocks. Mars is travelling through space after the Big Bang. Why is it So different to Earth? The crust on the sun's surface is a Solid. Then lave would be the next thing to follow. Looking at how things behave after an explosion we will see that they would not be separated but each object would of course contain both of these compounds. At a time in their outward journey they would of course be contained in the sun's orbital penny force. In compariSon Mars being the nearest to the sun would cool down quicker. Remember all the other matter is cooling down. You Also must fetch into this scenario the bible's story that the Lord brought together fire and Solids from space to create the Earth and the moon. In compariSon to the rest of the planets Earth has not been in space as long. Also before our lord decided on creating the Earth and the moon there was a larger gap between Mars and Jupiter.

But there was another thing that stuck out and that was it showed places where asteroids had hit the planet. Sorry about the misleading information I will just rectify it. If an asteroid had hit the planet it would not have shown that it had stuck out instead it would have shown a crater. One thing to notice about these craters is the size of them. Now please remember that this accrued billions of years ago. And the pieces that they found were only the size that could be held in the hand. I am not talking about a spoonful of matter that has been compressed by a black hole or what the strongest man on Earth could lift but a perSon of my abilities that has a problem lifting his head. If you have been taking notice of my previous revelations, then you will know that size of any asteroids that are going to hit the Earth are not going to be of a size that has no consequences to the destruction of our planet.

SPACE AND TIME

I have looked at various sites on the internet that are dealing with time and space. I am Sorry to say the time it took me I got lost in space. I could download this information. But not wanting to bore my readers that are like me they would find it impossible to take all this in.

The astronomers are finding scientific ways of learning how old different planets and asteroids are. I would think that there has been a lot of money spent by different departments in their research. With my method I could tell them the age of every planet sun moon and asteroids. You might be thunder- struck by my statement on the magnitude of this question. How can it be possible when all the boffins with all their money and equipment have taken So long? They use this equipment to date it by radiation decay. When you are dealing with time in the billions I suppose they could be out by a few million of miles. They have put the universes age at fifteen billion years. If they have used the radiation to find this out how can I dispute it. The physitrists and scientists have taken yonks to work this out I have taken minutes when I had to think about it. You might think that this is not possible. It was not done by magic it is what the astrologers have been saying from the beginning.

In physical cosmology, the age of the universe is the time elapsed since the Big Bang. The current measurement of the age of the universe is 13.799 ± 0.021 billion (10^9) years within the Lambda-CDM concordance model. The uncertainty has been narrowed down to 21 million years by the agreement of a number of scientific research projects, such as microwave background radiation measurements by the Planck satellite, the WilkinSon Microwave AniSotropy Probe and other probes. Measurements of the cosmic background radiation give the cooling time of the universe since the Big Bang, https://en.wikipedia.org/wiki/Age_of_the_universe-cite_note-arxiv-20121220-3 and measurements of the expansion rate of the universe can be used to calculate its approximate age by extrapolating backwards in time.

Explanation

The Lambda-CDM concordance model describes the evolution of the universe from a very uniform, hot, dense primordial state to its present state over a span of about 13.8 billion yearshttps://en.wikipedia.org/wiki/Age_of_the_universe-cite_note-4 of cosmological time. This model is well understood theoretically and strongly supported by recent high-precision astronomical observations such as WMAP. In contrast, theories of the origin of the primordial state remain very speculative. If one extrapolates the Lambda-CDM model backward from the earliest well-understood state, it quickly (within a small fraction of a second) reaches a singularity called the "Big Bang singularity". This singularity is not understood as having a physical significance in the usual sense, but it is convenient to quote times measured "since the Big Bang" even though they do not correspond to a physically measurable time. For example, "10^{-6} seconds after the Big Bang" is a well-defined era in the universe's evolution. If one referred to the same era as "13.8 billion years minus 10^{-6} seconds ago", the precision of the meaning would be lost because the minuscule latter time interval is swamped by uncertainty in the former.

Though the universe might in theory have a longer history, the International Astronomical Unionhttps://en.wikipedia.org/wiki/Age_of_the_universe-cite_note-5 presently use "age of the universe" to mean the duration of the Lambda-CDM expansion, or equivalently the elapsed time since the Big Bang in the current observable universe.

As can be seen in the above statements these are all based on theories. How is it with the billions of pounds spent on the equipment to find out how the universe works the different tests that are done come to a different time scale? I understand that when you are dealing with billions then being out by millions should not make that much difference. This I understand but when are they going to realise that there is far too much money spent on this subject. Anything that has happened in the universe over the last fifteen billion years there is not anything new going to happen. It would not surprise me that they will find Something else. I mean they have found black holes, string theory, space bending, gas gobbling giants, and there is probably life on the next planet they find.

It will make your mind boggle at my revelation of this information.

It will make your mind boggle at my revelation.
So how can the physitrists say that there are planets and sun's of a different age? They do this by radiation decaying the said material of planets and asteroids. I do not think they are looking at how the universe came about. It all started with the Big Bang. Does this not tell them anything? Is there not one physiatrist, scientist, mathematician or astromener able to work this out?

All matter in space is of the same age as it all came from the Big Bang
Meteoroid
From Wikipedia, the free encyclopedia
Not to be confused with meteorite.

From a meteoroid to a meteor and meteorite: how a meteoroid enters the atmosphere to become visible as a meteor and impact the Earth's surface as a meteorite.

A **meteoroid** (/ˈmiːtiərɔɪd/)[1] is a small rocky or metallic body in outer space.

Meteoroids are significantly smaller than asteroids, and range in size from small grains to 1 meter-wide objects.[2] Objects smaller than this are classified as micrometeoroids or space dust.[2][3][4] Most are fragments from comets or asteroids, whereas others are collision impact debris ejected from bodies such as the Moon or Mars.[5][6][7]

When a meteoroid, comet, or asteroid enters Earth's atmosphere at a speed typically in excess of 20 km/s (72,000 km/h; 45,000 mph), aerodynamic heating of that object produces a streak of light, both from the glowing object and the trail of glowing particles that it leaves in its wake. This phenomenon is called a meteor or "shooting star". A series of many meteors appearing seconds or minutes apart and appearing to originate from the same fixed point in the sky is called a meteor shower. If that object withstands ablation from its passage through the atmosphere as a meteor and impacts with the ground, it is then called a meteorite. An estimated 15,000 tonnes of meteoroids, micrometeoroids and different forms of space dust enter Earth's atmosphere each Asteroids, meteors, and meteorites ... It might be fair to say these rocks from space inspire both wonder and fear among us Earthlings. But knowing a bit more about each of them and how they differ may eliminate Some potential misgivings. While all these rocks originate from space, they have different names depending on their location — i.e. whether they are hurtling through space or hurtling through the atmosphere and impacting Earth's surface. In simplest terms here are the definitions:

ASTEROIDS

A large rocky body in space, in orbit around the Sun.

Meteoroid: much smaller rocks or particles in orbit around the Sun.

Meteor: If a meteoroid enters the Earth's atmosphere and vaporizes, it becomes a meteor, which is often called a shooting star.

Meteorite: If a small asteroid or large meteoroid survives its fiery passage through the Earth's atmosphere and lands on Earth's surface, it is then called a meteorite.

Another related term is bolide, which is a very bright meteor that often explodes in the atmosphere. This can Also be called a fireball. Asteroids are found mainly in the asteroid belt, between Mars and Jupiter. Sometimes their orbits get perturbed or altered and Some asteroids end up coming closer to the Sun, and therefore closer to Earth. In addition to the asteroid belt, however, there have been recent discussions among astronomers about the potential existence of large number asteroids in the Kuiper Belt and Oort Cloud. Asteroids are Sometimes referred to as minor planets or planetoids, but in general, they are rocky bodies that do not have an atmosphere. However, a few have their own moons. Our Solar System contains millions of asteroids, many of which are thought to be the shattered remnants of planetesimals – bodies within the young Sun's Solar nebula that never grew large enough to become planets.

The size of what classifies as an asteroid is not extremely well defined, as an asteroid can range from a few meters wide – like a boulder — to objects that are hundreds of kilometers in diameter. The largest asteroid is asteroid Ceres at about 952 km (592 miles) in diameter, and Ceres is So large that it is Also categorized as a dwarf planet.

Most asteroids are made of rock, but as we explore and learn more about them we know that Some are composed of metal, mostly nickel and iron. According to NASA, a small portion of the asteroid population may be burned-out comets whose ices have evaporated away and been blown off into space. Recently, astronomers have discovered Some asteroids that mimic comets in that gas and dust are emanating from them, and as we mentioned earlier, there appears to be many bodies with asteroid-like compositions but comet-like orbits.

GRAVITY

Every time you jump, you experience gravity. It pulls you back down to the ground. Without gravity, you'd float off into the atmosphere -- along with all of the other matter on Earth. You see gravity at work any time you drop a book, step on a scale or toss a ball up into the air. It's such a constant presence in our lives, we seldom marvel at the mystery of it -- but even with several well-received theories out there attempting to explain why a book falls to the ground (and at the same rate as a pebble or a couch, at that), they're still just theories. The mystery of gravity's pull is pretty much intact. So what do we know about gravity? We know that it causes any two objects in the universe to be drawn to one another. We know that gravity assisted in forming the universe, that it keeps the moon in orbit around the Earth, and that it can be harnessed for more mundane applications like gravity-powered motors gravity-powered lamps **Wrong**

All objects are kept in their respectable positions in our Solar system by the largest sun and that is in the Milky Way. If the matter was being drawn together then the universe would not be expanding. We know it is because the boffins keep telling us this. We in fact know this to be true. The reason being is that photos taken at different times of the universe shows that they are not in the same position but further apart. As for the science behind the action, we know that Isaac Newton defined gravity as a force -- one that attracts all objects to all other objects. We know that Albert Einstein said gravity is a result of the curvature of space-time. These two theories are the most common and widely held (if Somewhat incomplete) explanations of gravity. Although many people had already noted that gravity exists, Newton was the first to develop a cohesive explanation

WRONG

does not attract objects to other objects. Albert Einstein and Isaac Newton's revelation for how gravity works does not compute to the way I am looking at it. They are hoping Soon they will find an answer to how gravity works in space. Also they have given a reason that gravity is caused by the planets spinning. Now let me explain why objects of a different weight when dropped from the same height will reach the ground at the same time. This were Isaac Newton has got it right. And that is that objects of a different weight have an equal force pushing against it. To explain this to our beginners in how our world works is a ton weight has a force of a ton pushing up. all objects will react to the same principle.

If the experts find fault with my logics or theories what this means is this book is one long fairy tale.

Some religions don't eat bacon, others do not eat cows. Some eat bugs I couldn't eat bugs it would make my stomach crawl.

If you have taken it in what I have said about salt you will know that it is the only matter that was not part of the sun's elements. This is the only element on earth or anywhere else in the universe. It is the combination of Sodium and chloride. If our earth is going to last for jonks, then salt will be its demise. earth's seas will in time all finish up as a dead sea. This equals a vey large lump of salt.

www.ingramcontent.com/pod-product-compliance
Lightning Source LLC
Chambersburg PA
CBHW020733180526
45163CB00001B/215